T0123038

Science and Eastern Orthodoxy

MEDICINE, SCIENCE, AND RELIGION
IN HISTORICAL CONTEXT

Ronald L. Numbers, *Consulting Editor*

SCIENCE
and
EASTERN ORTHODOXY

From the Greek Fathers
to the Age of Globalization

EFTHYMIOS NICOLAIDIS

Translated by Susan Emanuel

The Johns Hopkins University Press
Baltimore

© 2011 The Johns Hopkins University Press
All rights reserved. Published 2011
Printed in the United States of America on acid-free paper
2 4 6 8 9 7 5 3 1

The Johns Hopkins University Press
2715 North Charles Street
Baltimore, Maryland 21218-4363
www.press.jhu.edu

Library of Congress Cataloging-in-Publication Data

Nikolaides, E.
Science and Eastern Orthodoxy : from the Greek fathers to the age of globalization /
Efthymios Nicolaidis.
p. cm.
Includes bibliographical references (p.   ) and index.
ISBN-13: 978-1-4214-0298-7 (hardcover : alk. paper)
ISBN-10: 1-4214-0298-x (hardcover : alk. paper)
1. Religion and science—History. 2. Orthodox Eastern Church—Doctrines. I. Title.
BX342.9.S35N55 2011
261.5′5—dc22   2011009987

A catalog record for this book is available from the British Library.

Index prepared by Nicholas A. Jacobson

*Special discounts are available for bulk purchases of this book. For more information,
please contact Special Sales at 410-516-6936 or specialsales@press.jhu.edu.*

The Johns Hopkins University Press uses environmentally friendly book materials,
including recycled text paper that is composed of at least 30 percent post-consumer
waste, whenever possible.

CONTENTS

Sometime around the year 49 CE, a Jewish tent maker and itinerant preacher, Paul of Tarsus, later known as Saint Paul, born in the first decade of the current era, paid a visit to Philippi in northern Greece. There he established the first Christian church on European soil. Paul had grown up in Tarsus, a largely Greek-speaking cultural center in present-day Turkey, near the Mediterranean Sea. As a youth he had studied in Jerusalem to become a rabbi and had joined the Phari-sees, a Jewish sect. Before converting to Christianity near the age of thirty, he had actively persecuted the growing band of Christians, as they were beginning to call themselves, who probably numbered under a thousand at the time. After his conversion, he spent considerable time in Damascus and Antioch in Syria (now Antakya in Turkey), at first proselytizing fellow Jews but later expanding his ministry to include Gentiles. After arriving in Greece about 49, Paul estab-lished Christian congregations in a number of cities, including Thessalonica and Corinth, where he stayed for a year and a half. According to Christian tradition, he died a martyr in Rome at about age sixty.

Despite suffering severe persecution at times, the Christians grew rapidly in number during the first three centuries after the death of Jesus. In 313 the Roman emperor Constantine (c. 272–337), who had embraced Christianity, issued an edict granting religious toleration throughout the empire. He also moved the capital of the empire from Rome to the Greek colony of Byzantium on the shores of the Bosphorus, separating Europe and Asia. He renamed it Constantinople.

By the fifth century, Christianity comprised five ecclesiastical jurisdictions, or patriarchates, centered in Rome, Constantinople, Alexandria, Antioch, and Jerusalem—with Rome claiming primacy. Already tension was growing between Latin Rome in the West and Greek Constantinople in the East. Within a cen-tury or two, the pope in Rome and the patriarch of Constantinople were scarcely talking to each other (and, in the mid-eleventh century, excommunicated each other). By the time Roman Catholic crusaders attacked their eastern brethren in 1204 and sacked Constantinople, the rupture was beyond repair.

Although the patriarch of Constantinople served as the titular head of the Eastern Orthodox Church, he did not acquire the power of the pope. Indeed, the emperors of the Eastern Roman Empire succeeded in controlling the patriarchate and in nominating the patriarch of their choice. The patriarch was officially elected by the ecclesiastical council, but the election had to be approved by the emperor, who in fact imposed his candidate. The emperors chaired the councils and succeeded in imposing their views on major issues, such as the presence of icons in the churches or a possible union with the Western Catholic Church. The Eastern Church was built on a hierarchical model. The patriarch controlled the metropolites, who controlled the bishops, who controlled the priests. Even the nomination of a metropolite had to be approved by the emperor. The monasteries elected their local leader, the *hēgoumenos*, but he remained under the control of the local metropolite.

As Christianity grew in the Greek-speaking world, science (or, more properly, philosophy) was in decline. For nearly a millennium, beginning with the early sixth century BCE, the Greeks, sometimes drawing inspiration from the Egyptians and Mesopotamians, had dominated the world of science. Following Pythagoran ideas of a world based on numbers and harmony, Plato formulated the idea that, as the most symmetrical body is the sphere, the cosmos must have a spherical shape and that the celestial movements must be circular and uniform. Soon after, Aristotle taught a coherent system that explained the phenomena in the heavens and on the earth as well as life forms. But even if his system aimed to explain everything, it did not become a dogma as it would later in the Middle Ages, and many other philosophical schools flourished. Epicurus (unlike Aristotle) believed in the existence of the void, Aristarchus of Samos proposed a system where the sun is the center of the world, and the Stoics believed that the universe conflagrates and is reborn eternally. By the third century BCE, the center of Greek scientific activity had shifted from Athens to Alexandria, Egypt, where an important mathematical and astronomical school flourished in affiliation with its Museum, which was a school plus a library, founded by the Greek dynasty of Ptolemy. Two great Alexandrian scholars, Euclid and Ptolemy, organized in a rigorous manner the geometrical and astronomical knowledge of antiquity.

With the growing dominance of the Roman Republic (later the Roman Empire) in the last centuries before the current era, scientific investigation occupied a far less important place in intellectual life than during the ancient Greek and Hellenistic periods, from the death of Alexander the Great to the Roman conquest of Constantinople. There were a few notable exceptions, such as the work carried out by the Alexandrian astronomer Ptolemy and by the anatomist Galen of Pergamum, both of whom were of Greek origin and lived in the second century CE. Although the Romans excelled in administration and engineering, they showed less interest in investigating nature than did the ancient Greeks. Thus, by the time Emperor Constantine converted to Christianity and the new religion became a dominant force in Greek culture, science had become a marginal activity.

The study of science and religion has recently become fashionable. In part, this has resulted from the unexpected resurgence of religion in recent decades, especially in its fundamentalist forms. After a long period during which most intellectuals saw the relationship between science and religion as a conflict symbolized by the trial of Galileo (perceived through the play by Bertolt Brecht), we now see this tendency changing, even reversing. People speak more and more of relations between religion and science that are not merely harmonious but privileged—and their focus now is not just Christianity.

Nevertheless, specialists have long remarked that if the subject is to be properly explored, then neither conflict nor harmony is an adequate generalization. Relations between science and religion are much more complex, and this is especially true if one examines the history of the relations between science and Christianity.[1]

In the context of this revival of interest, a series of scholarly books and articles has recently been published, trying to demonstrate a complexity that merely reflects the richness of civilizations, the richness of human history.[2] Local contexts, personalities, and political fluctuations are such that when one tries to generalize and to develop a model concerning relations between Christianity and science, one often falls into flagrant contradictions. Moreover, which Christianity is being referred to? The overwhelming majority of scholarly works follow the history of Western Christianity. The historiography that prevails in the Christian world, writes the archbishop of Canterbury, is that "the Church gradually builds up a centralized system of authority, filling the vacuum left by the fall of the Roman Empire; its ideological monopoly is challenged at the Reformation, and the map of the Christian world is reconfigured. . . . Even in some good and sophisticated surveys of world Christianity published in recent years, this remains the dominant picture. But Christianity is more various than this begins to suggest."[3]

The existing literature contains virtually nothing that pertains to the vast part of eastern Europe and western Asia that constituted the Eastern Empire, the *Byzantine* Empire. Here there was no political and institutional void; on the contrary, we observe an astonishing continuity that would be undermined only in 1204, the year of the conquest of Constantinople during the Crusades. This continuity also involved the perpetuation of an educational system put in place at the end of antiquity that was associated with both secular and ecclesiastical power, the latter placed under the authority of the former.

In the sciences, this translated into the perpetuation of a tradition of teaching ancient Greek science in the Greek language, sometimes as reviewed and modified by the Greek fathers of the school of Alexandria. Hence conflicts—when they existed—were not "science versus Christianity" but rather ecclesiastical conflicts that traversed the whole society and consequently also implicated the sciences by always coming back to the question of the importance of secular knowledge and the possibility for humankind to conceive of the Creation through science. For the medieval Oriental Empire, it was not a matter (as in the West) of rediscovering the ancients (who were always taught) but of finding equilibrium between secular knowledge and religious revelation and, hence, between science and Christianity. The Hesychast debate, which in the fourteenth century initially expressed the hostility of monks and lower clergy toward the secularization of the high clergy (meaning its alienation from the ascetic tradition), is an example. But here, too, lurk the preconceived ideas of dominant historiography. Hesychasts believed that a man through prayer and ascetics could have a vision of God and thus that true knowledge comes from this spiritual effort and not from acquiring secular knowledge. However, the ideological father of this movement, Gregory Palamas (1296–1359), based his ideas on the science of Aristotle and the geometry of Euclid in order to cogitate on locating the centers of the spheres of two elements, earth and water. What this movement seemed to be advocating was far from absolute hostility toward profane science. In effect, the Hesychast leader did not deny the utility of the sciences; he was more distrustful of the place granted to them by Byzantine power, seeing it as one of the causes of the secularization of high clergy.

The Orthodox Church as synonymous with Byzantium (and with "Byzantium after Byzantium") is often linked by historiography to a mystical approach that neglected profane knowledge.[4] Hence *science* and *Byzantium* were considered incompatible notions. However, recent work about the sciences in Byzantium, especially since the 1980s, has belied this approach and shown that the interface between science and Orthodoxy is as complex and varied as the analogous interface with Roman Catholicism in the West. In fact, the Byzantine world ultimately proved to be not only the guardian of a scientific tradition that comes down from antiquity but also receptive (albeit after debate) to the sciences of other civilizations, especially to those of Muslim cultures.

Concerning the long period from the fall of the Byzantine Empire to the end of the eighteenth century, historians of science have focused recently on the question of the reception by the Orthodox world of the "new European science." Here, too, the debates traversed the whole body of the Orthodox Church. In fact, the scholars of the Orthodox world who propagated this new science were in large part ecclesiastics, sometimes occupying high positions in the church hierarchy. Not until the rise of nationalism at the end of the eighteenth century and the effort of modernization in the nineteenth would there be forged in the Orthodox world the image of a man of science who was *not* linked to the church.

Only after the nineteenth century do we witness a discussion—sometimes turning to a debate—between science and Orthodoxy. Relations between the two reached the height of animosity a century later, during the communist period. After World War II, all countries with an Orthodox majority population—with the exception of Greece itself—were governed by communist regimes, and so generations of citizens were shaped by the dominant idea that Orthodoxy (as well as any religion) was incompatible with science. The Soviet Union's *Universal Encyclopedia of Youth* in its edition at the start of the 1960s—which is astonishingly rich in the history of science—unambiguously presents the position of the Christian Church during the Middle Ages: "In medieval Europe, there was no scientific development for several centuries. The best scientific achievements of Antiquity were forgotten. The Church savagely prosecuted any attempt at scientific explanation of the world."[5] As for Byzantium, apart from the reference to the geocentrism of certain church fathers and to a nonspherical world system held by Cosmas Indicopleustes, there is no mention of the sciences. By contrast, the medieval sciences of Asiatic peoples are richly presented, though obviously emanating from that part of Asia that belonged to the Soviet Union.

This doctrinal view was toppled after the political changes at the end of the 1980s, when many ex-communist scientists suddenly proclaimed themselves Orthodox and sometimes even favorable to Hesychast ideas. Meanwhile, historians of science, following international historiography on this point, concentrated on the study of relations between science and Orthodoxy by trying to grasp their complexity and variability in geographical time and space.

This book is the first general account of this story of the interrelationships between science and Christian Orthodoxy. As such, it is necessarily summary and partial. The book addresses the Orthodox Church as it was expressed by those councils recognized by the Patriarchate of Constantinople. This was the official church of the Oriental Roman Empire, Byzantium, which would be propagated in the Slavic world, including Russia. For want of space and expertise, I follow the history of science and Orthodoxy only partially with respect to its geographical scope. Thus, the book traces these relationships under the Byzantine Empire and, after its fall, in the Orthodox *millet* (communities) of the Ottoman Empire. It largely neglects the Slavic Orthodox Church, which followed its own course and deserves its own history.[6] For the period that followed the dismantling of the Ottoman Empire and gave rise to Orthodox nation-states in southeastern Europe during the nineteenth and twentieth centuries, I focus largely on the Greek example. Because the most compelling interactions involved cosmology and natural philosophy, I say little about medical and natural history, except in the last chapter, where I discuss Darwinism.

Although I have been working for a long time on the subject of science and Orthodoxy, the idea of writing a book about it—very short in comparison to the immensity of the subject matter—came from a dear colleague and friend, Ronald Numbers, who pointed out to me the absence of a book on the subject in the course of a lecture, "Science and Orthodoxy" at the University of Wisconsin in 2007. Without this encouragement, I would never have ventured to write a survey so ambitious in the time and space it aims to cover.[7]

Note on Transliteration. The current trend in translating Greek names is to transliterate them in a simplified orthography (for example, Georgios for Γεώργιος instead of George or Georgius or Geõrgios). This system has the advantage of being as close as possible to the original. Its major problem is its inconsistency with much of the existing literature, in which Greek names appear mostly in Latin or a vernacular language (for example, Basilius or Basil, not Basileios). In order to be as consistent as possible with the extant literature and to make reading easier, I have chosen to follow the traditional practice for ancient and Byzantine names (for example, George instead of Georgios, Aristarchus instead of Aristarchos). I have also followed conventional practice for the post-Byzantine period, where names are rendered by a simplified form of transliteration (for example, Theophilos, not Theophilus). In order to simplify typography, I have followed modern Greek orthography for the (very few) Greek citations.

Politics and Church
Science and Religion

312 (–337)	*Constantine the Great, emperor. Edictum Mediolani (religious freedom, 313)*
325	*First ecumenical council in Nicaea (condemnation of Arianism; caesaropapism)*
330	*Foundation of Constantinople*
c. 360	Themistius, director of the school in Constantinople financed by the emperor Constantius II
361 (–363)	*Julian, emperor: Reestablishment of paganism*
378	*Hexaemeron* of Saint Basil
379	*Apologia to Brother Peter on the Hexaemeron* of Gregory of Nyssa
381	*Second ecumenical council in Constantinople (Christianity official religion)*
386–388	*Homilies on Genesis* of John Chrysostom
395	*Division of the Roman Empire (Eastern-Western)*
415	Murder of Hypatia
425	Emperor Theodosius II founds the imperial university in Constantinople
431	*Third ecumenical council in Ephesus (condemnation of Nestorius)*
439	*First official Byzantine documents in Greek*
527 (–565)	*Justinian, emperor. Codex Justinianus (Byzantine law, 529)*
529	Justinian codex forbids pagan teaching in state schools
c. 550	Philoponus's *Creation of the World*. Cosmas Indicopleustes' *Christian Topography*

553	*Fifth ecumenical council in Constantinople: Condemnation of Origen and Theodore of Mopsuesta*
580 (–662)	Maximus the Confessor: Nonvalidity of secular knowledge
c. 600	*The emperor gives the title "ecumenical" to the patriarch of Constantinople*
610 (–641)	*Heraclius, emperor*
c. 630.	*Hexaemeron* of George Pisides
638	Encyclical against monophysiticism
642	*Arab conquest of Egypt*
c. 676 (–749)	John of Damascus, cleric and astrologer
680–681	*Sixth ecumenical council against monotheliticism*
717 (–741)	*Leon III, emperor*
726	*First iconoclast imperial act*
741	First proposed date for the end of the created world
753	*Iconoclast Council and measures against the monks*
775	Astrological predictions by Stephen the philosopher
c. 775 (–785)	Theophilos of Edessa, chief astrologer of Caliph al-Mahdī
787	*Seventh ecumenical council in Nicaea. Restoration of the icons*
802 (–811)	*Nicephorus I, emperor: Taxation of aristocracy and monasteries*
811 (–813)	*Michael I, emperor: Monasteries again received their privileges*
813 (–820)	*Leo V, emperor*
815	*Iconoclast Council and new measures against the icons*
c. 829 (–842)	Patriarch John Grammarian's activity in sciences and construction of palace automatons
842 (–867)	*Michael III, emperor*
843	*Council in Constantinople definitively restores the icons*
c. 855	Foundation of the School of Magnaura. Leo the mathematician at its head
858 (–864)	The scholar Photius named Patriarch. Photius's description of 121 secular books
864	*The Bulgarians become Christians: Beginning of the Christianization of the Rus*
867 (–886)	*Basil I, emperor: Restoration of the relations with the pope*
886 (–912)	*Leo VI the Wise, emperor: New laws to replace Codex Justinianus*

963 (–969)	*Nicephorus II Phocas, emperor: Prohibition against donating land and buildings to monasteries*
976 (–1025)	*Basil II Bulgaroktonos, emperor: Submission of Bulgarians to the empire*
c. 1000	Science of Islam influences Byzantine science
1008	Oldest preserved Byzantine *quadrivium*
1045	Michael Psellos, *hypatos* of philosophers
1054	*Schism between Eastern and Western Christianity*
1071	*Battle of Manzikert: Byzantine army defeated by the Seljuq Turks*
c. 1080	Simeon Seth court astrologer
1081 (–1118)	*Alexius I Komnenos, emperor*
1082	Trial of John Italos
1096	*First Crusade*
1107–1108	*Byzantine-Crusades wars*
1143 (–1180)	*Manuel I Komnenos, emperor*
1147–1149	*Second Crusade and Norman invasion in Byzantium*
c. 1160	First translations of Greek scientific texts into Latin in Sicily
1189–1191	*Third Crusade, conquest of Cyprus by Richard the Lionheart*
c. 1190	Teaching of philosophy at the Patriarchal School. Chair of *maistor* of philosophers
1204	*Conquest of Constantinople by the fourth Crusade; conquest of Peloponnese by the Crusaders*
1204–1205	*Empire of Trebizond. Despotate of Epirus; Empire of Nicaea: Theodore I Laskaris, emperor of Nicaea*
c. 1210	Theodore Eirinikos, *hypatos* of philosophers in Nicaea
c. 1226	The monk Nicephorus Blemmydes founds his school
1259	*Michael I Palaiologos, emperor of Nicaea*
c. 1261	Reopening of the imperial university. George Akropolites teaches mathematics
1261	*Reconquest of Constantinople by the Byzantines*
1262	*Foundation of the Despotate of Mystras*
c. 1275	George Pachymeres teaches science at the Patriarchal School
1300	Gregory Chioniades to Persia to study astronomy
c. 1300	*Quadrivium* of George Pachymeres, *Indian Calculation* of Maximos Planoudes

1589	*Foundation of the Patriarchate of Moscow and all Rus*
1622	Theophilus Korydaleus teaches Aristotle at the Patriarchal School
1632	Ioannis Kottounios succeeds Cremonini in Padua
1653	Greek Orthodox college in Padua
1661	Greek Orthodox college in Venice
1662	Paisios Ligarides in Moscow
1669	*Conquest of Crete by the Ottomans*
1675	Nicolas Spathar sent by the tsar to China
c. 1680	Ioannis Skylitzes presents the Copernican system
1683	*Ottoman army defeated at Vienna*
1685	Foundation of the Slavo-Greco-Latin Academy in Moscow
1692	Chrysanthos Notaras in Moscow
1700	Bishop Meletios Mitros presents the Copernican system
1716	Patriarch Chrysanthos Notaras's book *Introduction to Geography and the Sphere*
1723	Condemnation of Methodios Anthrakites by the church
1739	Chair of experimental physics in Padua influences Orthodox students
1753	Eugenios Voulgaris teaches the "new science" in Mount Athos
1760	Nikiphoros Theotokis teaches the "new science" at the Patriarchal School
1765	Iosipos Moisiodax teaches the "new science" at the Academy of Jassy
1769	*The Russians in Peloponnese. Uprising of the Christians against the Ottomans*
1774	*Küçük Kaynarca treaty between Russians and Ottomans gives privileges to Christians of the Ottoman Empire*
1776	Nikiphoros Theotokis in Russia
1790	*Anthology of Physics* of Rigas Feraios
1796	Sergios Makraios attacks the heliocentric system
1797	*The French occupy the Ionian Islands*
1808	Foundation of the Ionian Academy by the French and later the British
1809	*The Ionian Islands under British rule*
1813	*Serbian revolt*
1821	Patriarchal encyclical condemning science
1821	*Start of Greek national revolution*

1830	*Independence of Greece*
1833	*Otto of Bavaria, king of Greece*
1833	*Autonomy of Greek Orthodox Church*
1837	Foundation of Athens University supervised by the Greek Ministry of Education and Ecclesiastical Affairs
1864	*The Ionian Islands given to Greece by Great Britain*
1870	*Autonomy of Bulgarian Orthodox Church*
1873	First public lectures on evolution in Athens
1878	*Treaty of San Stefano recognizing Serbia and giving autonomy to Bulgaria*
1879	*Autonomy of Serbian Orthodox Church*
1890	The review *Anaplasis* attacks the Darwinists
1912	*Balkan wars*
1914	Trial of the Darwinist teachers in Nafplion
1922	*Defeat of Greek army in Asia Minor by the Turkish army*
1923	*End of the dismantling of the Ottoman Empire. Republic of Turkey: Kemal Ataturk, president*
1945	*Bulgaria and Yugoslavia under Soviet influence*
1946	*Declaration of the Christian Union of Scientists*
1946 (–1949)	*Greek Civil War*
1967 (–1974)	*Military dictatorship in Greece*
1980	*Greece, member of the European Union*

Science and Eastern Orthodoxy

CHAPTER ONE

The Activist and the Philosopher

The Hexaemerons of Basil and of Gregory of Nyssa

During the early centuries of Orthodoxy, discussions of nature focused almost exclusively on explaining the events associated with the six days of Creation described in the first chapter of Genesis. This interest resulted in a series of exegetical texts called the Hexaemerons, that is, the six days during which God created the world and living things. These commentaries sought to explain what God had wanted to reveal to Moses, the supposed author of Genesis, about his creation. Beyond these commentaries, there was little discussion of natural philosophy and cosmology, which had flourished among the ancient Greeks.

The Commentary on Genesis by Philo of Alexandria

The roots of the Hexaemeron tradition go back before the Christian era. The best-known non-Christian commentary on Genesis was written by Philo of Alexandria (c. 12 BCE–c. 54 CE) and called *De Opificio Mundi* (or *The Creation of the World according to Moses*).[1] It was the source of inspiration for later Christian texts, especially the Christian school of thought developed in Alexandria that tried to reconcile the Bible with Hellenic philosophy. Orthodox Jews considered Philo, a Hellenized Jew, as "assimilated to the Greeks," and yet Christians called him "Philo the Jew"; as for Greek philosophers, they had trouble acknowledging his monotheism founded on the Bible.

Philo's purpose was to convince pagan philosophers of the universality of Jewish monotheism. In his exegetical text, he interpreted the teaching of Moses in the book of Genesis through the mind of a Greek philosopher. According to Philo, the world (the cosmos of the ancients) is the creation of God and for this reason reflects the divine, as do humans who belong to this cosmos; both are images of God. "The world that is in law corresponds to the world and the world to the law, and a man who is obedient to the law, being, by so doing, a citizen of the world, arranges his actions with reference to the intention of nature, in harmony with which the whole universal world is regularized."[2]

According to Philo, order and the law in the Platonic sense (i.e., that order and law rule the cosmos and should also apply to society) argue in favor of Creation, because, if the world were not created, then it would be quite simply anarchic: "With regard to that which has not been created, there is no feeling of interest, as if it were his own, in the breast of him who has not created it. It is then a pernicious doctrine . . . to establish a system in this world such as anarchy is in a city."[3] But this created world, although marvelous—and this is the great difference between Jewish monotheism and Greek philosophy—cannot be compared to God. "Some men, admiring the world itself rather than the Creator of the world, have represented it as existing without any maker and eternal, and as impiously and falsely, have represented God as existing in a state of complete inactivity."[4] However, Philo's God is closer to Plato's god than to the Jews': "But Moses . . . was well aware that it is indispensable that in all existing things there must be an active cause, and a passive subject, and that the active cause is the intellect of the universe . . . while the passive subject is something inanimate and incapable of motion by any intrinsic power of its own, but having been set in motion, and fashioned and endowed with life by the intellect, it became transformed into that most perfect work, this world."[5] God is thus identified with the universal intellect, the active cause that acts in a constructive manner upon what is inanimate.

"Let no one ignorant of geometry enter": this motto of the Academy of Plato, on a par with the Platonic love of order and the law, is applicable to the cosmos of Philo. For according to the Jewish and Alexandrine philosopher, the world created by God is ordered mathematically and obeys the laws of Greek natural philosophy. Philo's God is a mathematician; he has created and ordered nature according to this science. It is numbers that explain the very duration of Creation. An omnipotent God would have created in a single instant the cosmos and everything included in it. That he spent six days doing so is precisely because of the order brought about by numbers. Drawing on the views of the Pythagoreans, who mathematized the world and who used the unit (the number one) to represent divinity, the number two for woman, and the number three for man, Philo wrote: "The things created required arrangement, and number is akin to arrangement, and of all numbers, six is, by the laws of nature, the most productive: for of all the numbers, from the unit upwards, it is the first perfect one, being made equal to its parts, and being made complete by them; the number three being half of it, and the number two a third of it, and the unit a sixth of it, and so to say, it is formed so as to be both male and female, and is made up of the power of both natures, for in existing things, the odd number is the male, and the even number is the female. Of odd numbers the first is the number three, and of even numbers the first is two, and the two numbers multiplied together make six."[6] This explains the duration of Creation. As for its beginning, Philo said that the first verse of Genesis—"In the beginning, God created the heavens and the earth"— refers not to time, for the latter is born along with creation, but rather to numbers: the number one, the monad, "should have been the first object created, being both the best of all created things and being also made of the purest substance."[7]

The created world, corporeal and perceptible, was copied from an intelligible and incorporeal world, also created by God, so that it could serve as an archetypal model. All the perceptible species of the corporeal world are replicas of the intelligible species of a primordial incorporeal world constituted by ideas. "That world which consists of ideas, it were impious in any degree to attempt [to know], but how it was created, we shall know if we take for our guide a certain image of the things which exist among us."[8] No other commentator on Genesis was ever as Platonic as Philo.[9]

Philo conceived of Creation as follows: on the first day, God created the world of ideas, and on the following days, he gave bodies to these ideas. An analogous line of thought would be maintained by several Christian fathers of the church, as well as by followers of the school of Alexandria such as Clement, Origen, Basil, and Dionysius, and also by those of the school of Antioch such as Theophilus of Antioch and Diodoros of Tarsus (about these schools, see chapter 2).[10] The difference is that according to the fathers, this pre-Creation does not mean a first incorporeal creation of material things. In fact, according to Basil, the angels will remain always incorporeal; moreover, they were created *before* the six days of the creation of the world. But according to Philo, the incorporeal world was created during these six days and would be transformed gradually into the material world: "In the first place therefore, from the model of the world, perceptible only by intellect, the Creator made an incorporeal heaven, and an invisible earth, and the form of air and of empty space: the former of which he called darkness, because air is black by nature;[11] and the other he called the abyss, for empty space is very deep and yawning with immense width. Then he created the incorporeal substance of water and of air, and above all he spread light, being the seventh thing made; and this again was incorporeal and a model of the sun, perceptible only to intellect, and of all the light-giving stars, which are destined to stand together in heaven."[12] Then come the ideas of all the future perceptible things it according to the theory of sacred numbers. Early Christians, such as Origen or Basil, who spoke of cosmology and the philosophy of nature—hence, of science—read Philo attentively; although they developed different ideas, they were all strongly influenced by this Hellenized Jewish philosopher.[23]

Christian Hexaemerons

Christian commentaries on the genesis of the world began with Origen (c. 185– c. 254), a Christian scholar who taught in Alexandria and gave his *Homilies on Genesis* in Caesarea in the decade of the 240s CE.[24] As was the custom at the time, tachygraphic scribes recorded these spoken homilies. These *scholia* were very different from those of Philo, moving away from exegesis founded on numbers and entering into spiritual allegory. Origen found for each phrase of Genesis a parallel between the body and the soul. But because of his condemnation by the church (owing to his conflict with the bishop of Alexandria, Demetrius, and his posthumous characterization as a "heretic" by the councils of 545 and 553), future generations regarded his ideas with suspicion. Thus, his writings did not enjoy widespread dissemination in the Greek Orthodox world, and the original Greek *Homilies on Genesis* was lost. In fact, we know the work only through a later Latin translation from the early fifth century.

The fourth century was crucial for the formation of Christian dogma. It witnessed the first ecumenical synods, or councils, which normalized local religious practices and decided among different interpretations of scriptures. The earliest was organized by the emperor Constantine in the town of Nicaea in 325 in response to religious debates that threatened the domestic peace of the Roman Empire. By this act, the Christian Church was integrated into state institutions. The fourth century also saw the fight against the notion that Jesus was not of the same substance as God, which gave rise to the dogma of the Trinity—as well as the emergence of liturgical religious practices and celebrations, such as Easter.

The Hexaemeron that most influenced the Christian conception of nature was the one composed by Basil of Caesarea (c. 330–79), the son of a well-off Christian family from in the region of Pontus. Born in Caesarea in Cappadocia, Basil was educated by his grandmother, Macrina, and his father, also called Basil. With his friend Gregory of Nazianzus, he studied in Constantinople and then Athens, where he became familiar with Greek philosophy, including natural philosophy, the antecedent of science. Returning to Cappadocia around 355, he practiced law and taught rhetoric but soon abandoned them and devoted himself to religion. In 370 he became bishop of Caesarea. Basil played an important role in the formulation of Christian dogma and contributed greatly to Christian liturgy and to discussions of dogma bearing on the substance of God and the Trinity. He wrote major dogmatic, ascetic, and exegetical texts on various aspects of social and cultural life. Over time, Orthodox Christians came to consider him the first church father.[25]

Basil's *Nine Homilies on the Hexaemeron*, written in 378, is the exegetical text on Genesis that had the greatest dissemination in the Eastern Christian world and was the source of inspiration for most later exegetical texts, whether from Eastern or Western Christendom. Like other such works, which were numerous at the time, it celebrated the world as God's creation. Basil's homilies are, as their name indicates, sermons that he delivered in the church of Caesarea before a diverse audience, as described by his brother, Gregory of Nyssa:

> He spoke to a large audience in this church and made provision for them to receive his message. Among the many listeners were some who grasped his loftier words whereas others could not follow the more subtle train of this thought. Here were people involved with private affairs, skilled craftsmen, women not trained in such matters, together with youths with time on their hands; all were captivated by his words, were easily persuaded, led by visible creation and guided to know the Creator of all things.[26]

In addressing such a group, composed of Christians of Caesarea of all ages and social classes, the bishop had to keep his message simple and didactic; thus, he used accessible allegories and avoided high-flown philosophical concepts.

Like many other towns in fourth-century Asia Minor, Caesarea fostered Greek culture. Debates concerning questions of philosophy before a wide public must have been a common thing, and the depiction of discussions in Alexandria described by Cosmas Indicopleustes (see chapter 2) almost two centuries later was no doubt valid for Caesarea in Basil's day. The art was in the popularization, in order to convince the great majority of humble folk to whom Gregory spoke, while also retaining some lofty arguments for educated people who were present. According to the same source, Basil had "to be understood by the crowd but also admired by the best."[27]

Like those of Origen, the homilies of Basil were copied down as they were delivered for the purpose of their dissemination. But because of their dogmatic importance, it is highly probable that they were carefully prepared before being delivered and then revised before being distributed in written form. As the fruit not of improvisation but of reflective composition, they responded to questions of dogma concerning the Creation and what follows from it, including matter, time, space, life, relations between the created world and God's world, and the nature of evil. By giving these homilies at a mature age (he was forty-eight and had already written almost all of his works), Basil wished simultaneously to incite the public to lift its gaze from the created world to the Creator, to codify a story (Genesis) that was foreign to the tradition of Greek cosmology and make it concordant with the image of the world of his intellectual milieu, and finally to combat the "external" enemies of Christianity (pagans, Manichaeans) as well as the "internal" enemies (adepts of Arianism and Christian Gnosticism).[28]

From their publication, Basil's homilies on the Hexaemeron aroused a storm among pagan philosophers, at the time still numerous and powerful. These philosophers found Basil's theses unfounded because they were in flagrant contradiction with science. We should not forget the continued existence at this time of two of the most famous pagan philosophical "schools" of antiquity; the school of Athens was still active, and Alexandria's would not become Christian until two centuries later. Moreover, the era of the reign of Emperor Julian the Apostate (361–63), who would try to restore pagan religion, was not far off.

The commentaries that the pagan philosophers wrote to refute Basil's theses have not come down to us, but we know their essence through the response of Gregory of Nyssa to these critics. During the summer of 379, a year after Basil had given his homilies and shortly after his death, his brother Gregory wrote his *Apologia to His Brother Peter on the Hexaemeron*. Peter (349–92), bishop of Sebastia and younger brother of Basil and Gregory, appears to have asked Gregory to complete Basil's unfinished work, for he had planned to follow up his homilies on the Hexaemeron with a commentary on the creation of man according to the premise that the world was created in order to receive him. Over Easter 379, Gregory wrote homilies (probably composed to be published) entitled *The Creation of Man*, and shortly afterward, *Explanation of the Hexaemeron*.

Little is known about Gregory's life; he was born around 335 and died sometime after 394. Tradition tells us his brother Basil named him bishop of Nyssa in 371 against his will. He is considered the most scholarly of Greek church fathers, a fine connoisseur of philosophy, a brilliant and prolific writer, but a mediocre orator. His writings were circulated much less than Basil's. Despite the important role he played against Arianism at the ecumenical council of 381, the Eastern Church considered him a philosopher rather than a theologian.

In contrast to the *Hexaemeron* of Basil, the *Apologia* of Gregory is a text addressed solely to scholars; following the concepts of science of the day, he responds to the arguments of these scholars against Basil's Christian natural philosophy. Basil and Gregory were fervent Christians and, at the same time, scholarly men, permeated with Greek culture and, hence, with science. Their image of the world could not help being that of the philosophers of the Eastern Roman Empire of the fourth century—or at least those who had not made a special study of astronomy.

At this time, the astronomy of Ptolemy (second century BCE) was well established. His cosmological system (as explained in the *Hypotheses of the Planets*) and his mathematical solutions for the movements of heavenly bodies, as detailed in the *Grand Mathematical Syntax (Almagest)* and in his astronomical tables assembled in the *Handy Tables*, had become dogma for the mathematicians of the Roman Empire. Commentaries explaining his work had already appeared and would soon be eclipsed by those of Theon of Alexandria, which were based on the course of astronomy he taught at the Museum around 370. This allows us to suppose that in the fourth century Ptolemaic astronomy was taught in summary fashion in the curriculum that followed the *egkyklios paideia* (secondary school), which included arithmetic, music (harmony), geometry, and astronomy.

The causes of the Ptolemaic success that lasted at least fourteen centuries have been abundantly discussed by both astronomers and historians. From the astronomical viewpoint, his complete solutions to planetary movements, which were based on eccentric circles and epicycles (geometric tools that were already well known to the mathematicians of his day); his admirable solution of the equant circle; and, finally, his explicit, systematic, and didactic presentation of all the celestial movements, all combine to explain this success.[29] From the cosmological viewpoint, two elements helped perpetuate his system: first, the dogma developed by Plato that all celestial motions had to be uniform circular movements, which became the touchstone of Greek cosmological thinking; and, second, the dogma of geocentrism, which would be fervently adopted by Christian cosmology.

So then, what was that summary image of the Ptolemaic world, the cosmos of the Greeks that Basil and Gregory knew from their studies? First, it was a spherical universe that contained spherical celestial bodies with an immobile spherical earth at the center of the world, extremely small (like a point) compared to the dimensions of this world, as determined by the sphere of stars ("fixed" to differentiate them from "planets," a word that might be translated as "those that wander"). Next, this cosmos included seven planetary spheres (in order: the moon, Venus, Mercury, the sun, Mars, Jupiter, and Saturn), plus an eighth that carried fixed stars and marked the boundary of the universe. What existed beyond this sphere is a mysterious nowhere. A single movement was permitted for the heavens, the uniform circular motion that was the natural movement of heavenly bodies that thereby move without constraint. The apparently complicated movements of planets (as celestial phenomena) were explained by the activation of various uniform circular movements, with several at work for each planet. The essential characteristic of Hellenic astronomy was this uniform circular movement of the heavens—not geocentrism. Since Plato, all mathematicians had accepted this movement, although a minority of them did not place the earth at the center of the world.

It is difficult for us to determine how much Basil and Gregory knew about mathematical astronomy. Basil's line of argument in favor of a world that must change according to the conditions of the life that is promised to our souls meant he must attack any secular wisdom that is "so keenly aware of vain matters," and in doing so, he incidentally lets us glimpse his evident knowledge of what astronomers do:

> They who measure the distances of the stars and register both those in the
> north, which are always shining above the horizon, and those which lie about
> the south pole visible to the eye of man there, but unknown to us; who also
> divide the northern zone and the zodiac into numberless spaces; who carefully
> observe the rising of the stars, their fixed positions, their descent, their recur-
> rence, and the length of time in which each of the wandering stars completes
> its orbit.[30]

In another passage, Basil seems to know the difference between the sidereal day (the time it takes for the earth to make one rotation on its axis) and the solar day (the time between appearances of the sun).[31] And so we may suppose that Basil's and Gregory's studies must have included an elementary treatise on astronomy, such as the *Introduction to Phenomena* by Geminus (first century BCE). We may imagine the treatises used by students who were not going to follow higher studies in astronomy, thanks to a textbook of the following century that has come down to us, *On the Sphere* by Proclus (412–85). This manual for students presents elementary notions of the geometry of a sphere in order to teach them the divisions into circles of the terrestrial and heavenly spheres. Basil's and Gregory's knowledge of mathematics probably did not go beyond this type of treatise. Moreover, Basil had studied in Athens during a period when the teaching there centered on philosophy (including the philosophy of nature); someone who wished to do advanced mathematics and astronomy would go instead to Alexandria.

The knowledge of physics of the two fathers of Cappadocia was clearly more solid than their knowledge of astronomy. Gregory appears to have been an expert on the physics of Aristotle and the Stoics; Basil also knew about Stoical physics, having probably read one or more of the lost textbooks by Poseidonius (135–51 BCE), the celebrated Stoic, founder of the school of Rhodes, and author of numerous treatises on physics, mathematics, and astronomy that must have been well known in the fourth century.[32]

In any event, the image of the world that seems to have prevailed in Greek teaching of the fourth century was rather Aristotelian: the world was divided in two by the sphere of the moon. Underneath this sphere, the world called sublunary was subject to corruption and change; it was composed of four basic elements: earth, water, air, and fire. Each of these elements had a natural place and tended to return to it as soon as it was displaced. Earth and water, the heavy elements, tended toward the center of the universe, and the earth, being heavier, moved with greater speed. Air and fire, light elements, tended toward the lower boundary of the lunar sphere; fire, being lighter and hence more rapid, is placed underneath the air. Above this sphere, the so-called superlunary world, was incorruptible and eternal; it was composed of an immutable element, ether or *pemptousia* (the fifth element) that is always found moving in circles. In addition, since for Aristotle a void cannot exist in the cosmos, the sublunary region was filled with the four elements and the superlunary region with ether.

But then how could these facts be reconciled with the cosmology that underpins the book of Genesis? Nowhere in the Bible's first book is there any allusion to this spherical earth that is found suspended, alone, at the center of a spherical universe. The biblical world is closer to the Eastern cosmologies consisting of a rather flat earth, floating on the water, and overlaid with heaven in the form of a dome.

The Limits of Reason

Basil's challenge was to perceive the Bible armed with the tenets of Hellenic science, a very different task from that of the Greek philosophers:

> An appropriate beginning for one who intends to speak about the formation
> of the world is to place first in the narration the source of the orderly arrange-
> ment of visible things. For, the creation of the heavens and earth must be
> handed down, not as having happened spontaneously, as some have imagined,
> but has having its origin from God. What ear is worthy to hear the sublimity
> of this narrative? How should that soul be prepared for the hearing of such
> stupendous wonders: cleansed from the passions of the flesh, undarkened by

The Darkness, the Nature of Light, Day and Night

Having excoriated the Gnostic interpretation of formlessness, Basil attacked the Manichaean interpretation of darkness, which according to the Bible covered the abyss. For the Manichaeans, darkness was the force of evil fighting against God, who represented Light. Thus, darkness has its own existence that preexists Creation. Since this preexistence—and also the existence of the forces of evil not created by God—contradicts his omnipotence, Basil maintained that darkness did not have its own existence but simply signified an absence—"it is a condition incident to the air because of the deprivation of light."[55] When the heavens "were made by the command of God, surrounding completely the space enclosed by their own circumference with an unbroken body capable of separating the parts within from those outside, necessarily they made the regions within dark, since they had cut off the rays of light from the outside."[56] According to Basil, God's world was inundated with light, which illuminated the created world. This light has infinite speed: "Such is its nature of ether, so rare and transparent, that the light passing through it needs no interval of time."[57] Here we recognize the influence of Aristotle, who rejected the void, because it would allow, in the absence of friction, bodies to move at an infinite speed.

The infinite speed of light "passes our glances along instantaneously to the objects at which we are looking."[58] But at great distances, the grandeur of objects contracts, for our visual power does not manage to traverse the intermediate space, "since the power of sight is not able to cover the space between but is exhausted in the middle and only a little part at a time reaches the visible objects."[59] Thus, Basil's theory of vision had light being the support for vision, but it is a ray that emanates from the eye to bear upon objects. So this theory of vision appears confused and contradictory, which was not uncommon in antiquity. Basil mingled influences from Plato, Hipparchus, and the Neoplatonist Plotinus.[60]

According to Gregory (who always followed Hellenic philosophy more closely than did Basil), the light of the created world is not the same as the light of God's world; it was created along with matter and emanates from it. Immediately after the Creation, light, which is only pure fire, was hidden in the molecules of primordial matter that we referred to above. "Fire spreading everywhere was obscured by the excess of matter. But since its force[61] is rapid and mobile, when the signal for the genesis of the world was given by God to the nature of beings, the heaviest matter appeared and immediately everything was illuminated by light."[62] According to Gregory, this light-fire was freed from primordial matter at the divine command, "Let there be light."

Immediately after the lighting of the world,[63] God created the day and the night by separating the light from the darkness. In Basil's geocentric system, the existence of a day and a night before the creation of a sun that orbits around the earth required an explanation. He said this phenomenon was created by the very movement "when that first created light was diffused and again drawn in according to the measure ordained by God, when day came and night succeeded."[64] This time was set at twenty-four hours, very exactly measured by "the return of the heavens from one point to the same point once more,"[65] which informs us about Basil's astronomical awareness, for he seems to know that twenty-four hours (one revolution of the fixed or star sphere) corresponds to the duration of the sidereal and not to the solar day.[66]

Gregory's explanation was quite different, inspired by Aristotelian physics: since light is only pure fire, it must be weightless and naturally moves upward, and, like any element, its parts have a tendency to gather together. After the Creation, it "projected from heterogeneous elements like an arrow and, having a natural and light ascendant movement, it advanced in the universe; having traversed material nature, it could not continue in a straight line, for spiritual nature cannot be mixed with the material, and fire is material."[67] Fire is therefore obliged to adopt a circular movement at the boundary of the sphere that delimits the created world. And so day and night were created by the earth that, as a heavy body, is found at the center of the world.

The Separation of Waters and the Ninth Celestial Sphere

The sentence in Genesis, "and God said 'Let there be a firmament in the midst of the waters, and let it separate the waters from the waters'" (1:6) caused great difficulty for theologians who were expert in the Hellenic world system, for it brought a new and inexplicable element into their astronomy. This account of creation, along with the passage in the Psalms about "the heaven of heavens," would entail a multiplicity of spheres above that of the fixed stars, which would characterize the cosmology of the Middle Ages.

According to Basil, the firmament is a different sphere from this one, constructed at the moment of Creation for the purpose of separating the created world from God's world. The firmament is a sphere created below the sphere delimiting the cosmos, and it is identified with the sphere of fixed stars; it is made up of a more solid matter (hence its name),[68] and its purpose is to contain the heavenly waters of which the Bible speaks. The latter are of the same nature as earthly waters and have no allegorical meaning: "We have also some argument with the birth of Christ. In Basil's cosmos, the sunlight was replaced by light coming from God, and the sun-star is only the support for this light: "Now this solar body has been made ready to be a vehicle for that first-created light."[81] In his argument for the sun-as-vehicle of light, we can find proof of the clear distinction he made between *qualities* and *essences* (ουσια) of bodies: "Do not let what has been said seem to anyone to be beyond belief, namely, that the brilliance of the light is one thing and the body subjected to light another In the first place, we divide all composite bodies into the recipient substance and the supervenient quality."[82]

In his French edition of the *Hexaemeron*, Stanislas Giet says that Basil admitted that the moon sheds borrowed light.[83] I think that, on the contrary, Basil dismissed the opinion (which was well established in the educated Greek world of his day) that the moon reflects solar light, because he did not want to grant the least primacy to the sun. The phases of the moon, the central argument against its having its own light, are explained simply by the difference between light and support: "For although it wanes and decreases, it is not consumed in its entire body, but, putting aside and again assuming the surrounding light, it gives us the impression of diminishing and increasing. . . . If you observe it when the air is clear and free from all mist . . . it is possible for you to see its dark and unlit part circumscribed by such a circle." Later on: "Do not tell me that the light of the moon is brought in from the outside because it decreases when it approaches the sun but increases again when it moves away."[84]

The same desire to underestimate the role of the sun led Basil to comment on the phrase, "And God made the two great lights." *Great* did not signify the best quality but rather the most spacious; all the stars are constructed of the same matter, and they are the supports of the same light. The difference in brilliance is due only to their dimensions and their distance from the earth. The sun's dimensions are enormous in relation to the earth, for it has the same apparent diameter seen from any point on earth.[85] Its orbit is also very great, for it shines on the whole earth at the same time.[86] By contrast, the dimensions of stars are very small, since, "though the stars in the heavens are countless in number, the light contributed by them is not sufficient to dispel the gloom of night."[87]

Meanwhile, Gregory's material theory of light confounded light/fire and luminous bodies. But this fire that has emanated from primordial matter was not a unique and pure element: it was divided into eight sub-elements that had slightly distinct qualities; they had time to separate in the three days between the creation of light and the creation of luminous bodies. Because light comes from fire, it obeys Aristotelian laws of nature, for fire, a light element, obeys laws analogous to those of heavy elements. In the same way that liquids differentiate themselves and are located according to their weights (Gregory gave the example of oil, water, and mercury), so the various kinds of fire are differentiated and located at various heights according to their lightness.[88] "All the adequate and homogeneous molecules of light adhered to each other according to their kinship and were distinguished from the heterogeneous."[89]

According to this physics, the lightest and fastest homogeneous kinds of fire were concentrated at the boundary of the real universe to form the multitude of stars,[90] which are in perpetual movement without moving from their respective place on the sphere of the fixed, for the nature of their order is immobile, but their individual nature is mobile.[91] These stars are composed of homogeneous fires, but the physical nature[92] of each of these fires differs such as to confer on them a precise place on the fixed sphere: "[This nature permits them] to be placed at the middle, or at the south or north, or to occupy an intermediate space, and to form the galaxy[93] or zodiac circle, and within this circle to form some composition of stars."[94] Later on, the planets were placed according to the nature of their fires: the less this nature was rapid (οξυτατη κινησις), the lower this star was placed, ending with the moon, which was the most material and the most dense of celestial bodies.[95] Gregory did not make any commentary on the planets as such; however, he did remark that the Bible (under the general name "stars") mentions the five other planets.[96]

According to Gregory, the sun and moon are larger than the other stars, but their luminosity is not due to their dimensions but depends on the nature of the fire inherent in each celestial body. The sun's luminous nature is much less than that of the fixed stars (in fact, it is located in a much lower circle than the fixed), and its apparent luminosity is due to its greater proximity to the earth than theirs. Light is effectively spent by distance. The dimension of the sphere of the fixed is so enormous[97] that we can consider the sun to be at the center of the world.[98] In the wake of Ptolemaic astronomy, Gregory not only considered that the earth is "like a point in relation to the sphere of the fixed" but thought the orbit of the sun can also be considered as a point in relation to the orbit surrounding the real universe, here taking up the idea of Aristarchus of Samos. This reminds us of the famous passage of Archimedes in the *Psammites* that presents the heliocentric system of Aristarchus and indicates the infinite relation between the orbit of the earth around the sun and the sphere of the fixed. In this small book, Archimedes said that, according to Aristarchus, "the sphere of the fixed stars, situated about the same center as the Sun, is so great that the circle in which he supposes the Earth to revolve bears such a proportion to the distance of the fixed stars as the center of a sphere bears to its surface."[99]

The moon, said Gregory, "orbits a place in proximity to the earth and its nature may be considered as average, for it partakes as much of the non-luminous force as of the luminous force. In effect, the density of its essence has weakened its brightness, but with the reverberation of the rays of the sun, it is not completely alienated from luminous nature."[100] Contrary to Basil, Gregory was a partisan of the reflection of the sun's rays by the moon, while supporting the existence of specifically lunar light, albeit very weak.

The Laws of Nature

In his commentary on the separation of waters, Basil presented his ideas on the laws of nature, which were later taken up by several Byzantine ecclesiastical scholars. For the Aristotelians, what was later called the laws of nature were qualities linked to bodies: water flows because the action of flowing is innate to this element; heavy bodies fall because their natural place is at the center of the world. However, Basil thought these "qualities" of bodies were not innate to them from their creation. It was the will of God that dictated the behavior of these bodies, and thus there is some independence between laws and elements. Water did not necessarily have the property of flowing before God gave the commandment, "Let the waters be gathered into one place." "Why does Scripture reduce to a command of the Creator that tendency to flow downward which belongs naturally to water? Because as long as the water happens to be lying on a level surface, it is stable, since it has no place to flow; but when it finds some incline, immediately . . . it is borne naturally down slopes and into hollows. . . . Reflect that the voice of God makes nature, and the command given at that time to creation provided the future course of action for the creatures."[101]

By the time of his death, Basil, bishop of Caesarea, the first among the founding fathers of Orthodox dogma, intransigent combatant against the pagans but also "heretic" Christians, a brilliant orator, prolific writer—in short, an activist of the Christian Church—had laid the foundations of the Christian conception of nature and consequently defined the relations between science and faith. His affirmations were incisive, for "the truth is one," and it does not like contradiction. One year later, his brother Gregory returned to several of Basil's theses, which, when he found them in contradiction with the scientific knowledge of the era, he put them on the right path of natural philosophy. Although the Byzantines asserted they were the heirs of Greek science, this claim would be based on the theses of Basil and not those of Gregory, who, although recognized as a great philosopher, would never acquire the status of a great theologian in the Eastern Church.

Two Conceptions of the World

The Schools of Antioch and Alexandria

Historians have identified two currents of biblical interpretation among the early fathers of Eastern Orthodoxy: the school of Alexandria and the school of Antioch. The former was developed in the second and third centuries by Clement of Alexandria (c. 150–c. 215) and his student Origen, and leaned toward an allegorical interpretation that left great freedom of thought in reading biblical texts. The latter arose in the third and fourth centuries in the theological school of Antioch; its leaders included Lucian of Antioch (c. 240–312), Diodoros of Tarsus (d. c. 390), John Chrysostom (c. 350–407), and Theodore of Mopsuesta (c. 352–428). It preferred a literal interpretation that left little place for philosophical investigations.

Despite their sometimes literal reading of the Bible, the Cappadocean fathers Basil of Caesarea and Gregory of Nyssa belonged to the school of Alexandria. Their interpretation of Genesis, as we have seen, incorporated a system of the world that came from Greek and Hellenistic culture: a geocentric universe in the form of a sphere and a spherical earth.[1] In contrast, the Hexaemerons associated with the school of Antioch relied on a system of the world coming from Asiatic cultures: universes shaped in diverse forms and a flat earth. Throughout the Middle Ages, Orthodox scholars tended to embrace the cosmology associated with the school of Alexandria. The leading Byzantine mathematicians and philosophers adopted it, as did the royal court and most of the patriarchs. By contrast, the cosmology of the school of Antioch became the popular cosmology of the Middle Ages, propagated both orally and in such written works as the "Lives of the Saints."[2]

John Chrysostom

John Chrysostom, whose name signifies "golden mouth" in Greek, was a great orator. Born in Antioch around 350, son of a prosperous family, he studied at the theological school of Antioch with Diodoros of Tarsus and at an early age devoted himself to theology. He became archbishop of Constantinople in 397, but his religious fanaticism and his militant asceticism earned him many enemies at the Byzantine court in Constantinople; consequently, he was deposed and died in exile in 407. In 438 Emperor Theodosius brought his remains back to Constantinople; thereafter, he would be considered as one of the founding fathers of Orthodox dogma.

Between 386 and 388, John delivered in the cathedral of Antioch seventy-eight homilies on Genesis. Of course, many of these discourses were sermons dealing with moral and theological questions that do not always have much to do with cosmology. But in a few passages John developed cosmological themes or a philosophy of nature that present a different image of the world from Basil's.

In order to respond, as Basil did, to the Manichaeans, John began by supporting the fundamental thesis of Christian cosmology: matter did not exist before the Creation.[3] But in order to explain the first verse, "In the beginning, God created the heaven and the earth," he departed from the Cappadocean fathers to claim that God first created heaven and then laid out the earth underneath it.[4] He created the roof first and then the foundations, for he was capable of doing something that men could never do. John's cosmology was simple: a flat earth covered by a single heaven in the form of a vault. Heaven is immobile; it is the stars that move, and their movement serves to determine time.[5] We are far from the universe of the Cappadocean fathers, which we recall was composed of several heavenly spheres (which by their movement entrain the stars) and a spherical earth.

John believed that men should not excessively value the earth because it is not on account of *its* nature but rather by the will of God that they enjoy its bounty. It is for this reason that God wished the earth to be at the outset invisible and unformed, in order to demonstrate that it is he who offers its blessings. The causes of the invisible and unformed are clearly indicated by Moses when he says that "there was darkness on the surface of the void, and the spirit[6] of God moved above the waters." This quite simply means that the earth was covered by the abyss of waters (hence, its form did not appear) and that everything was in darkness, for the light had not yet been created (hence, invisibility), an explanation quite similar to Basil's.[7]

Cosmas and Philoponus clashed around 550, with the forthcoming council in view. The polemic was said to have started in the streets of Alexandria, where Cosmas gave public lectures to support his views, even demonstrating them with the help of experiments. This debate produced two very important documents for the relations between science and the Christian religion: Cosmas's *Christian Topography* and Philoponus's *Creation of the World*.

Cosmas had not received a systematic education. He considered as his teacher an adept of the Nestorian School of Nisibis in Persia, an important home for teaching the doctrines of Theodore of Mopsuesta in the sixth century. This teacher was the *katholikos* (archbishop) of the Persian church, Mar Aba, whom he had met in Alexandria. But Cosmas had also read a great deal.[17] Returning from his voyages, he composed four books: *Book on the Stars' Courses*, *Christian Topography*, *Commentary on the Song of Songs*, and a book of geography, of which only the first has come down to us, owing to its importance in the religious debate. This book perfectly illustrated the idea of the cosmos (and consequently of the profane sciences) among adepts of the Antioch school; many of these ideas would linger in nonscholarly orthodox milieux until the seventeenth century.

The Christian Topography of Cosmas

Cosmas was clear: one should not trust the arguments of "our own reason to understand the form and position of the universe," for this was to "mock" "divine Scripture as a whole as [if they were] myths" and to characterize "Moses, the prophets, the Lord Christ, and the apostles as speech-makers and impostors."[18] This was true not only for pagans but also for those Christians who wanted "to accept *both* Christian *and* pagan principles."[19] And what is the greatest deception of the science based on reason? According to Cosmas, it was to attribute to heaven a spherical form and a circular movement.

Basing his work on the Nestorian theory of two conditions (the divine and the human, which coexist in Christ), Cosmas conceived of the world in two kinds of space (see figure 1). The first extends from the earth to the firmament, "this world here, in which are found angels, men, and the whole present condition." The second extends from the firmament to the vault and is the world "in which Christ, at his ascension, entered first, inaugurating for us a new and living path."[20] What is the form of this world? It has nothing to do with a sphere; rather, the world is in the image of a tabernacle, a parallelepiped (or cube) topped by a vault.

The parallelepiped is our perceptible world and has at its base the earth, which is longer from west to east than from north to south; its roof is the firmament, surmounted by celestial waters. In the middle of the earth lies a continent, the interior earth or *oikoumenē*, surrounded by a single ocean. This ocean is itself surrounded with a band of solid earth, the "earth beyond," the habitation of men before the Flood. Along the east side of this earth lies Eden, earthly paradise. Four rivers flow out of this paradise, of which the Nile, passing through the ocean, flows into the *oikoumenē* and then into the Mediterranean ("Romaic Gulf").

As for heaven, "it is fastened by its own extremities to the extremities of the earth along its four sides, forming the quadrangular shape of a sort of cube; in its upper part the heaven is rounded into a belt in the lengthwise direction and seems to be a vast vault."[21] As we have noted, heaven is divided into two by the firmament. This whole edifice is suspended in the void, for it is founded "upon itself and not on something else."[22]

We find in Cosmas's text all the popular imagery of the Middle Ages for earth and heaven: the inhabited earth beyond, which is the support of heaven; the terrestrial paradise (to get there, you have to cross the uninhabited ocean and earth); the kingdom of heaven above the firmament, where the righteous will go to be with Christ; and the image of the flat earth and the walls that hold up the sky.

How did he explain the day and the night in such a world? The surface of the earth is not horizontal but has an elevation in its northwest part. This is why, in navigating toward the southeast, ships go faster than in the opposite direction; for the same reason, the waters of the Tigris flow faster than those of the Nile. This elevation in the north seems like a mountain; when the sun, which turns in our parallelepiped world, passes behind this mountain, this makes night in the inhabited lands and day in the uninhabited Nordic regions of *oikoumenē* and in the earth beyond the North. According to Cosmas, travelers coming back from the lands of the great North had seen the sun never setting during the night and thus verified this thesis.[23] The rising and setting of stars occur for the same reason. The eclipses of the moon are produced by the shadow of this earthly elevation, not by the shadow of the spherical earth as the Greek astronomers said. As for solar eclipses, they probably result from the shadow of the moon, as Greek astronomers maintained, which is not in fact incompatible with Cosmas's system.[24]

Creation and Angels

Cosmas stated that God created on the first day, in a single stroke, the heaven and the earth in the form of a house, containing the mixture of water, air, and fire in public that the whole Greek cosmological system was false, for in this system, the sun was much larger than the earth. But in Cosmas's system, the sun hiding behind the northern elevation of the earth was much smaller, and the earth served as base for the whole universe. So Cosmas had to show the Alexandrians in a public experiment the relative size of the sun.

In the Alexandrian context, Cosmas could not ignore the methodology of Greek science, and here he was in contradiction with his own beliefs. In effect, to counter his opponents, he proceeded as a philosopher, by demonstrating with the aid of experiment and reason the falseness of the arguments of the partisans of Hellenic science, his adversaries. Here is his argument. According to optics, "when the illuminating body is large and the illuminated body is small and both are spherical, then the shadow formed must necessarily be a cone."[34] Hence, the shadow of a small sphere produced by the sun would necessarily take the shape of a cone, and the size of the projection of the sphere would diminish with distance. Under the Alexandrian sun, which produced very clear shadows, Cosmas hammered a nail into a small wooden sphere, held it up, and—miraculously— the shadow was not conical, but the size of its projection was (at least) equal to the dimensions of the sphere. This purportedly demonstrated the falseness of the optics of Greek philosophers.

Here we see the level of Cosmas's education. He was not ignorant of science; he must have read (or learned by oral means) the principles of optics. But, as would be the case for any half-educated person, the conclusions that he drew from his knowledge were often erroneous. Cosmas could not understand the value of the angle of the shadow produced by the illuminating body when it was at a great distance from the illuminated body, which in fact gives a nearly parallel shadow. And he was not in a position to know the laws of light diffusion in the atmosphere, which give the shadow a blurry contour, making it appear in our case larger than the illuminated body. Similarly, the demonstrations that he gave with the aid of drawings to demonstrate the error of the Greek theory of "climates" and to determine the size of the sun, although they were based on elementary principles of optics, also led to erroneous results.

According to the ancient Greeks, the northern hemisphere is divided into seven climates, that is to say, zones between terrestrial parallels. The annual movement of the sun around the earth in a circle that is inclined in relation to the earthly equator (the ecliptic) produces the variation in the length of the sun's shadow at midday in the course of seasons. Evidently at the same moment, at different climates, the size of the shadow is different. But, according to Cosmas, all this could not happen in the framework of a spherical earth and of a sun larger than it, and he strained to demonstrate this with the help of a drawing in which a sun larger than the earth does not produce parallel rays on the latter. On the contrary, a small sun lighting a large flat earth produces, according to the placement, a shadow of a different value (see figure 2), which was what he wanted to demonstrate.

Philoponus's *Creation of the World*

According to Philoponus, the book of Genesis, as its name indicates, tells us about the creation of the world by God. Moses's goal was to show that "the world did not come into existence in a spontaneous manner" and that "it is not of a superior and divine substance"; he was not giving a dissertation on nature.[35] Such dissertations would be given by later physicists (those who study nature, *physis*), who asked the following questions: What is the material principle of things, and is there one principle or several? What are they, if several? What is the nature of celestial objects; are they different from objects situated under the moon, and is their substance corruptible?[36]

Most of these "physicists" maintained contradictory theories, or theories contradictory with the reality; nevertheless, some of them, inspired by Moses, did advance our knowledge of nature. Plato, "the flower of philosophy," had imitated Moses in his tale of the origin of the world in the *Timaeus*.[37] The most famous astronomers of antiquity, Hipparchus and Ptolemy, had been inspired by the existence, according to Moses, of the firmament found underneath the sphere delimiting the world, and conjectured a ninth sphere without stars, external to all the rest, which would imprint the movement of the world from west to east, an explanation of the displacement of the equinoxes.[38] Philoponus was clear that "the goal of the present treatise is to show as much as possible that nothing in the cosmogony of the prophet is in disagreement with the order of the universe; on the contrary, the causal explanations given later by physicists find on numerous points their origin in the writings of Moses."[39] A literal reading of the Bible contains many pitfalls. In this case, one would have to accept that the dwelling of God is heaven, and the earth is a stepping-stone—but that is inadmissible, for God does not need a dwelling; moreover, Philoponus added ironically, "let's go see where God lived before He made heaven . . . and what the stepping-stone was before He made the earth."[40]

The Beginning

By following Basil, and taking up his linguistic argument drawn in turn from Aquila's translation of the Bible (see chapter 1), Philoponus affirmed that matter, space, and time were created instantaneously. According to Greek philosophers, time is engendered by the movement of heaven; they thought that time was the measure of the circular movement of celestial bodies. Philoponus cited Plato's sentence in the *Timaeus:* "Time, then, and heaven came into being at the same instant in order that, having been created together, if ever there was to be a dissolution of them, they might be dissolved together."[41] Hence, before heaven came into being, time could not exist.

Against the Antiochians, Philoponus affirmed that the earth and heaven were created simultaneously. At the same time, between the two extremes (the center and the limit of the cosmos), which are the element earth and the element that composes heaven, there were created the three other elements that compose the world: water, air, and fire. This explains the phrase "and darkness covered the face of the abyss," for the abyss is the deep water that covered the earth, and the darkness is the air deprived of light (according to most Greek physicists, air deprived of light is black).[42] As for fire, according to the physics of Aristotle, it is contiguous with air and spreads right up to the internal surface of heaven (sphere of the moon), and for this reason Moses does not mention it explicitly. This fire element is not the flame that we know, but the "combustible" that serves to enflame and from which are born lightning, shooting stars, comets, and all the sublunary phenomena according to Aristotle.[43]

Angels

Apart from articulating the physics of the perceptible world, Philoponus's goal was to discuss the question of how angels were created in order to refute Cosmas and the partisans of the Antiochians. Here he proceeded by reasoning from the absurd, while at the same time basing his arguments on physics and theology.

If angels were created at the same time as the cosmos, then they are like living beings, either corporeal or incorporeal but attached to a body. If they are corporeal, they are composed of either the four elements or else a fifth imperishable element, such as the Aristotelian ether. In the former case, they would be perishable, an impossible thing; in either case, they would be deprived of reason, because matter cannot engender reason. If they are incorporeal but attached to a body, this material body could have been created only along with the material world. But Moses relates this creation in so much detail that he surely would have mentioned such a fact. If not, he would have considered the angels less important than the animals (souls attached to a body)—an impossible thing.[44]

Therefore angels are incorporeal, that is to say, their place is not found in the cosmos, as claimed by Theodore of Mopsuesta and his adept Cosmas. For place is a dimension; hence, "to be in a place pertains only to bodies that are extensive in three dimensions,"[45] that is, to corporeal beings.

If angels were not created *during* the creation of the perceptible world, then they were created either before or afterward. The latter case is impossible, for according to scripture they praise God just after the sixth day of Creation, and moreover they are much more important than material and corporeal beings like the animals, for example, because they were created after them. Consequently, there remains only the possibility that the angels were created *before* the perceptible world, Q.E.D.

The Shape of the Cosmos

Philoponus's perception of the world is Aristotelian, but with a biblical terminology and contributions from Ptolemaic astronomy. According to Aristotle, the spheres of the planets are homocentric, whereas Ptolemy spoke of exocentric circles and epicycles bearing the planets. In addition, Ptolemy, following Hipparchus, added a ninth sphere without stars for the revolution of the heaven from west to east every thirty-six thousand years, responsible for the precession of the equinoxes. This sphere (according to Philoponus) is the first heaven. Indeed, in the image of the Trinity, Philoponus believed that the world is divided into three:

1. The first heaven, spherical in form, without stars, that delimits the world
2. The second heaven, called the firmament by analogy with Aristotle's superlunary world
3. The sublunary world of Aristotle

Spherical in form, the second heaven carries the celestial bodies. It is divided into eight spheres (one for each planet plus one for the fixed stars). Either the planetary spheres are exocentric or they bear epicycles. The firmament moves as a whole at one revolution per day, and this movement engenders the diurnal revolution of celestial bodies. The other spheres also move and are responsible for the apparent movements of planets. This world is immutable but not incorruptible and its duration is equal to that of time as a whole.[46]

Philoponus explained that Moses (who was talking to people who were not all educated) sometimes calls heaven the "space between the earth and the moon," because the air and fire above our heads are perceived as heaven.

What is this world made of? For the sublunary world, this is simple: the four elements, earth, water, air, and fire. Even Plato and Aristotle would agree on that. Things get complicated when we consider the matter of heaven.

First, if we follow Aristotle, heaven has to be composed of some kind of matter, for "in effect, nothing of what exists is ever empty."[47] But as concerns this matter, Philoponus distanced himself from Aristotelian physics, which had a clear and precise explanation for the physics of the whole cosmos and argued for the existence of a fifth incorruptible element whose natural movement is circular. This physics would prevail in the Middle Ages, eclipsing that of Plato, who wanted only four elements for the whole cosmos, and who thought that heaven was composed of these same elements and that stellar matter was essentially fire. Philoponus, although he followed Plato by rejecting the fifth element of Aristotle, did not give a precise and definitive explanation but launched into conjunctures.[48] Because Moses very often calls air *water*, then by proclaiming that "God said 'Let there be a firmament in the midst of the waters,' the prophet tell us that the second heaven is essentially made of air and water. This accords with its transparence, for air and water are transparent (whereas earth and fire are not), and with the solidity of the firmament for air and water, they can be transformed into more solid substances (according to Philoponus, air is humid in nature, hence it can solidify)."[49] However, this affirmation should be moderated, Philoponus said, when conjecturing "according to phenomena, the demonstrations of Plato, and what Moses leads us to think."[50] Later on, Philoponus was more confused when he spoke of the matter *between* the two heavens (the first heaven and the firmament): he declared that Moses designates with the homonym *water* the matter that is between the first heaven and the firmament, and that this matter "is of a very subtle substance . . . either aqueous or aerial, or something combustible, or of some other nature."[51]

Thus, Philoponus launched an explanation of celestial waters that is quite different from all his predecessors: the waters below the firmament are air plus the water of which the second (superlunary) heaven is composed, but they are also a part of the sublunary world, whereas the waters above the firmament (celestial) are made of the element of which the first heaven is composed. These "waters are separated in two by places and essences, having in common only the name; God created the firmament as a frontier."[52]

Philoponus categorically insisted that the world moves circularly, contrary to the beliefs of the Antiochians, who argued for an immobile world. Moreover, he was a fervent partisan of Ptolemy, who "rejected with contempt all the hypotheses of the ancients [e.g., Aristotle] and whose conceptions are the most evident and the most concise."[53] But Philoponus must not have read Ptolemy completely, or else he must have had a good reason to present a very simplified version of the great astronomer's system. The stars' movements, he said, are due to nine spheres, of which some are exocentric. Ptolemy's system is far more complicated than this. In order to account for the complicated apparent movements of the planets, Ptolemy used up to six spheres per planet.

Although Philoponus gave quite a few details of mathematical astronomy (the Ptolemaic values for the duration of the revolutions of planets, the sphere of the fixed, and the ninth sphere), his essential purpose was to demonstrate the sphericity of the world and the earth, as well as the movement of spheres, in order to refute the Antiochians.[54] Evidence of the earth's sphericity follows from the teaching of Moses himself, who stipulates (according to Philoponus) that the earth was initially entirely covered with water. But for this to take place it would be necessary for its form to be spherical. Indeed, according to Aristotle's physics, the natural place of the elements earth and water is at the center of the world. The element earth, being heavier, tends to occupy the very center in a symmetrical way, that is, in the form of a sphere. Water tends to cover earth's sphere also in a symmetrical way, in a form of a sphere of larger diameter. According to Philoponus, the fact that earth was entirely covered by water necessitates this theory and so demonstrates earth's sphericity. Later, God shaped earth's sphere in such a manner as to make the land appear.[55]

To maintain that it is the spheres that move and not celestial bodies, he demonstrated that the sphere of the fixed moves as a whole; if it did not, the stars would need variable speeds in order to remain in their respective places. He explained the eclipses of the moon and sun by basing his arguments on the Ptolemaic system.[56]

Following Basil, who condemned astrology for being a forecast of human destiny, Philoponus declared that "the practice of astrology is hated by God and is ruinous for the soul in that it distances those who indulge in it from God and from the hope that we place in Him, and prevents them from asking Him by their prayers and by a good life what is suitable for them, on the pretext that nothing happens outside Destiny."[57] The stars can indeed serve as signs for meteorological predictions but not for the lives of humans.[58] He went so far as to make a clear distinction between astronomy and astrology, by defining the rigorous observations of astronomy as compared to unfounded astrology: the former are the observations that make us "understand exactly the stars' movements."[59]

He adopted the tradition of Philo that the stars and sun and moon were created later than what grows in the earth so that men might not venerate them as gods.[60] For the same reason, the stars should not be confused with light; they are merely the medium for it, because light is a simple quality, whereas the stars are composed of a body and a quality. Just as glowing things emit different lights according to their species, so the light emitted by all stars is not the same; this is due not to the nature of light (which is unique) but to the composition of their bodies. Thus, each star emits its own light, even the moon, which emits a very feeble light visible at lunar eclipses, although its luminosity is specifically due to the reflected light of the sun.[61] In this way, Philoponus moved away from Basil, who believed that the light of the stars was the same as the light of God's world, and from Gregory, who believed that light has different natures. He returned to Plato, who believed that the stars are made essentially from fire and that the light is a quality of fire.

In the sublunary world, the Creation followed an order that went from the most simple to the most complicated or, more precisely, from less perfect beings to the most perfect.

First were created the elements "as principles of all the composite bodies, animate and inanimate bodies." Philoponus recognized three faculties in animate beings: "the faculties of feeding, growing, and engendering a being similar to itself."[62] It is evident that the elements themselves are inanimate.

Then the first animate beings were created, the plants, which, although they possess the faculties described above, have neither sensibility nor movement.

Aquatic animals were the second animate beings to be created, because water is the second element after the earth in the Aristotelian hierarchy (earth, then water, air, and fire). They are more perfect than the plants (they have the faculties of movement and sensibility) but less perfect than aerial beings, for water, being thicker than air, lets in less light, sound, and odor and, hence, diminishes the sensations of aquatic beings in relation to aerial ones.[63]

In third place came the birds (air is the third element). They are more perfect than the fish, for their senses and imaginations are more acute, but they are mostly wild.

Fourth came the terrestrial animals, still more perfect, for some of them (e.g., domestic animals) are sensible.

Last came "the most perfect of all the animals, man."[64] And here is how man is born: first, sperm falls into the womb, and there it receives the vegetative life; in growing, it receives animate life and becomes a living being; finally, still in the womb, it receives the rational soul. A soul, according to Aristotle, is specific to the bodies of animals. The *rational* soul, according to Philoponus, is thus found attached to bodies. The soul of man is made of a substance superior to (and more divine than) that which composes bodies (the four elements); hence, it is neither on the basis of the same substance nor at the same time that the body and the soul are created. The soul of man is an intelligible and reasonable substance, with life breathed into it by God after the creation of bodies, and so it is totally different from that of animals, whose souls were engendered by the earth ("let the earth produce a living soul"). The human body is made of the four elements, despite what is literally said in scripture, for without water, clay cannot be fashioned; also, "it is evident" that we partake of air and fire. Philoponus cites Homer in support of his thesis, saying of dead bodies, "All went back to earth and water."[65]

Note that Philoponus, in line with the knowledge of his era, maintained that the birth of certain species can take place without sperm; it can be spontaneous, as can be observed when the dirty and infertile earth is covered with greenery after irrigation. This results from the fact that "from the beginning, God deposited in the inanimate elements the spermatic reasons of future beings."[66]

Originating from a theological debate over the interpretation of the Bible, two conceptions of the world confronted each other in the streets of Alexandria in 550. On one side (Philoponus's) lies a world that we can explain, within the limits that God imposes on us, by profane knowledge, the natural philosophy of the ancients, and in which the phenomena of nature are due to the principles of the physics imposed at the Creation. On the other side (Cosmas's) lies a world whose reality is based on theological symbolism and in which the natural philosophy of the ancients is replaced by the mechanics of angels, responsible for all the phenomena of nature. Both Philoponus and Cosmas were considered later as heretics, the first for his belief that the Trinity is not of a single essence, and the second because he was a follower of Theodore of Mopsuesta. Nevertheless, this did not prevent their cosmological views from being spread in the Byzantine Middle Ages. Philoponus's views gained the approbation of the philosophers and continued to be discussed by Byzantine scholars, whereas Cosmas's views enjoyed a certain popularity among the monks and the lower clergy. Their popularity among uneducated people is attested by similar views developed in the "lives of the saints," which were common in later Byzantine literature. In these texts, the savant saints, answering questions on nature, involve the angels in natural phenomena and present earth as a flat surface.

CHAPTER THREE

No Icons, No Science

The End of a Tradition?

Natural philosophy may have flourished in the sixth century, especially at the famous school of Alexandria, but the tradition of cultivating Hellenic scientific knowledge quickly came to an end. Various factors, many not yet fully studied, led by the end of that century to this reversal. Thereafter, Christianity, the dominant ideology of the Eastern Roman Empire, would demonstrate almost total lack of interest in natural philosophy. Eastern schools quit teaching secular science and copying related manuscripts. Within a single generation, knowledgeable scholars virtually disappeared. Thus began the Middle Ages in the Roman Empire of the East, which historians would later call Byzantium.

Despite a sometimes delicate relationship between Christianity and secular knowledge, Byzantine authorities had not hindered the teaching of the sciences before the start of the sixth century. The schools in Constantinople that had developed in the fourth century not only taught natural philosophy but hired pagan professors to do so. The best-known school was financed by Emperor Constantius II and directed by the pagan Themistius (317–c. 390), a distinguished philosopher. The same emperor in 357 financed a center for copying manuscripts, including ancient texts on the sciences. In such ways, imperial power sustained, both financially and institutionally, structures in which science was cultivated and transmitted with a view to training future civil servants of the empire.

In the fourth century, the principal enemies of triumphant Christianity were no longer pagans outside the church but heretics inside it. Constantius himself had come under the influence of the Arians, and his primary concern was for peace within Christianity, not the persecution of pagans. Moreover, as we have seen, during the same period important men of the church such as Basil and Gregory of Nyssa gave their consent to Greek scientific knowledge. In the following century, in 425, Emperor Theodosius II (whose wife Athenaïs-Eudokia, daughter of an Athenian *rhētor*, was known for her love of science) promulgated a decree organizing "the imperial university" of Constantinople, the *auditorium*, in which philosophy (even the philosophy of nature) was taught. Reports from around 475 claiming that the imperial library contained 120,000 books may have exaggerated the size of the collection, but even a smaller library shows the good intentions of imperial Christian power toward scholarship.[1]

The reversal of this favorable situation for the sciences began during the reign of Emperor Justinian, from 527 to 565. Justinian and his wife, Theodora, were of humble origin and surrounded themselves with men of the same class; they displayed little interest in secular knowledge and quarreled with the aristocrats who did. Unlike Constantius, Justinian was not an intellectual, and he had not pursued studies during his youth as was common among the sons of the aristocracy. Wanting to reform the empire, he promulgated the famous Codex Justinianus in 529, which stipulated that "those who do not follow the catholic and apostolic church and the orthodox faith," meaning heretics, Jews, and pagans, were not authorized to become civil servants of the state. Consequently, they could no longer, under cover of any form of teaching whatsoever, induce good souls into error. Did this directive go so far as to forbid the teaching of pagan knowledge, hence of secular science? We saw in the preceding chapter that this was not at all the case in Alexandria, where Philoponus, a Christian, continued to teach Greek science. Some historians have maintained that Philoponus published his *Against Proclus* in 529 precisely in order to declare his adherence to Christianity and so continue to enjoy the favor of the imperial court.[2] But things were different for the school of Athens: testimony speaks of the interdiction in 529 of the very teaching of philosophy.[3] More generally, it appears that Justinian decided not to pay the salaries of science teachers, who had been formerly paid by the state to teach in the towns of the empire. Although the probable reason for this refusal to pay them was that Justinian needed revenue to finance his prestigious construction projects, such as the Church of Saint Sophia, the result was disastrous for the sciences. Without the prestige of education financed by the state, they were quickly dismissed by the dominant Christian ideology as useless pagan knowledge and abandoned in favor of endless theological discussions, which culminated in the debate over icons in the eighth century.

We do not know the immediate impact of Justinian's policy on the university in Constantinople. The history of this university remains to be written, since reliable sources that would allow us to follow the highs and lows of this institution are lacking—or, more precisely, are not certain. Was it really closed by Emperor Phocas (r. 602–10), as some scholars have claimed, and reopened by Emperor Heraclius (r. 610–41), during whose reign it enjoyed a new (but brief) revival?

Moreover, the distinction between the patriarchal school (probably called *oikoumenikon didaskaleion*) and the imperial university is not clear through the centuries. One thing is certain: from the end of the sixth century until the start of the ninth, scholars were scarce. But, although neglected in comparison with the preceding period, the sciences were not totally abandoned.[4]

Among those who did have a relation with the sciences, the best known was Stephen of Alexandria, a philosopher and astronomer who held the title ecumenical philosopher during the reign of Heraclius in the seventh century and who taught the works of Aristotle, Plato, Hippocrates, and Galen, in addition to the *quadrivium* at the *oikoumenikon didaskaleion* of Constantinople. Stephen, who wrote a text on Ptolemaic astronomy, was familiar with the mathematical sciences of the ancients. His activities and other sources suggest that a scaled-down "program of instruction" did survive, dispensed by both private tutors and organized schools. This instruction promoted basic literacy, using books of psalms instead of Homer to teach children to read; it also offered a course of higher studies, the *egkyklios*, which consisted of secular science, the *trivium*, and the *quadrivium*, the later being the main course of scientific education.[5]

Study of the *quadrivium* (arithmetic, geometry, music, and astronomy) goes back to late antiquity and took a definitive form around the fifth century. The Greek translation for the word *quadrivium*, *tetraktys*, appeared around the eighth century and is found in the contemporary lives of saints.

A *quadrivium* written in 1008 is the oldest Byzantine one that has survived, and it contains a range of topics: in arithmetic, the elementary theory of numbers, powers, and analogies and also the qualities of numbers based on the classification of Nicomachus; in geometry, Euclid's notions of *The Elements* (elementary trigonometry, planes, and solids) as well as references to the famous problems unsolved in antiquity, such as the Delos problem (if we double the volume of a cube, what does its side measure?). In music, there were the Pythagorean analogies; in astronomy, notions of spherical geometry, calculation of dates by the Egyptian calendar (indispensable for following Ptolemy), the sun's movement and its eclipses (only the principle and not the means of calculation), the same for the moon, notions of the movement of planets with a summary description of their epicycles, the constellations, notions from Ptolemy—mentioning that the stars can predict natural phenomena but not human actions. Pythagoras, Nicomachus of Gerasa, Archimedes, Euclid, Theon of Alexandria, Ptolemy—these are the sources of the treatise that (without demonstrating any originality) constituted a solid basis for advancing in the study of mathematics by using the texts of Diophantus or in the study of astronomy by using Theon's commentary on the *Almagest* or (rarely) the *Almagest* itself. The anonymous author of the 1008 *quadrivium* constantly repeated that faith precedes science, no doubt in an effort to get his book approved by the church.[6]

George Pisides and Maximus the Confessor

George Pisides was a well-known poet during the reign of Heraclius. We know that he was deacon in the church of Saint Sophia in Constantinople. A man close to the court and to the patriarch, he composed a Hexaemeron in verse.[7] This long poem (1,889 lines) was in fact a hymn to creation, but Pisides praised nature rather than God himself. In fact, he hardly addressed the Genesis story of the six days of Creation, as traditional Hexaemerons did; his work described the world and its living beings in the manner of very ancient poems on nature that were little known in the Byzantine period.

As a man of the church and at the same time a scholar, Pisides belonged to the tradition of the Alexandrian school: his earth is spherical and at the center of the world, surrounded by the vast vault of the spherical heaven in perpetual movement.[8] Writing only a century after the Neoplatonist scholar Proclus, who had argued for the eternity of the world, Pisides devoted some thirty lines to Proclus in his poem, summoning him to be humble before God's creation of the world.[9]

In effect, Pisides summarized in literary form the knowledge and fundamental conceptions of the world taught by the school of Alexandria, which were far removed from the philosophical erudition of the first Hexaemerons. The largest part of the poem is quite simply a description of animals and the earth's rarities: Pisides composed a small encyclopedia of the marvels of Creation, from the pelican to the hippocampus, rather than try to describe the process of creation in the terms of natural philosophy.

Pisides was much appreciated as a poet, and his *Hexaemeron* enjoyed some success in the Orthodox world—and (interestingly) would later become well known in western Europe thanks to numerous editions, including a translation into Latin after 1584. No doubt its success derived from its verse form. Although written by a man of the church, this *Hexaemeron* is closer to a secular form of literature than to a religious one. Dedicated to the patriarch Sergios, Emperor Heraclius's right-hand man, and doubtless with his approval, this poem was well appreciated by the church's hierarchy and the court. It was a return to forms of expression of the Roman period, which coincided with a brief turnaround during the reign of Heraclius in favor of the sciences of Greek antiquity. In this text, we also find one of the rare mentions in the Byzantine era of scientific instruments. The Byzantines were too theoretical to be concerned with describing observations made with the aid of instruments, to the point that some historians wonder if they ever even observed the eclipses they had predicted. Pisides mentioned the dioptra (an instrument for determining angles) and its scope but not in order to discuss observations. He spoke of it in an allegorical manner, to show off the intellectual capacities of the patriarch.[10]

From the end of the sixth century onward, scientific education was hindered more by a dismissive mentality toward science among some emperors and patriarchs than by a policy of outright hostility. Maximus the Confessor (580–662) was a celebrated theologian, guardian of Orthodoxy, but also a supporter of Pope Martin in Rome, the enemy of Emperor Constans II, which caused him to be tortured and to have his tongue and right hand cut off. Maximus did not believe that the knowledge acquired by experience was valid, since our senses deceive us. Sensations, said this prolific author, were part of the irrational, and therefore they belonged to the animal part of the soul. It is intelligence that enables people to perceive reality.[11] Maximus distinguished between soul (ψυχή) and intellect (νους). Animals, like humans, have a soul, but only humans possess intellect. But sensations are a part of the soul that is inferior to intellect, and it is with the latter that we approach God and, hence, true knowledge. Maximus, although inspired by Gregory of Nyssa, was above all an ascetic and a mystic; he especially sought union with God. Profane knowledge, although not rejected implicitly, was of little interest to him.[12]

The Iconoclasts and the Sciences

Maximus's lack of interest in knowledge obtained by the observation of nature and his belief that the sole worth of knowledge was whatever brought humans closer to God were shared by Orthodox believers of the seventh and eighth centuries, although with some variants. Orthodox mysticism was a current that would take various forms across these centuries. What changed was the method of obtaining this spiritual elevation to the divine. From those who practiced an ascetic way of life to the extreme *stylitēs* who lived on the top of columns in order to be closer to the sky, mysticism had a great influence among monks, whose main concern was to feel God's presence, thus neglecting secular knowledge. Sometimes dominant and sometimes marginal, mysticism never ceased to occupy an important place in this church. During the iconoclast period, this ideology became dominant; hand in hand with state policy, it would deal a decisive blow to scientific learning.

What is uncontested is the lack of texts relating to the sciences between the end of the sixth and the start of the ninth century. However, during this same period a great revolution took place, one comparable to that of the discovery of printing—namely, the discovery of cursive writing, which would enormously facilitate the copying of texts and make manuscripts much less expensive. This new form of writing was much faster and, by taking up less space, allowed savings in the precious support materials, whether paper or parchment. This contradiction between the revolution brought by writing in cursive and the disappearance of nonreligious manuscripts can be explained only by the abandonment of copying centers in the monasteries and evidently by a disdainful policy toward profane knowledge, which had until then furnished a great number of texts to copy.[13]

The debate over icons during the so-called iconoclastic period found its origins in the Asiatic part of the empire, and it culminated at the start of the eighth century, absorbing all the imperial energies. Probably influenced by the Muslims, whose rapid expansion in the Mediterranean posed a great threat for Byzantium during this period, Orthodox theologians launched an important and ferocious polemic over icons: What did they represent? Should they be considered as sacred objects, capable of effecting miracles? Should artists represent God? Very soon the debate moved from theological milieux to involve the whole society; henceforth patriarchs, emperors, and officers of the empire would be either *iconoclasts* (who smashed icons) or *iconodules* (who worshiped icons).

When a whole society debates a theological issue so passionately, it is likely to disregard secular knowledge. And the sciences, less useful to the march of the empire than the law, for example, were the first to be neglected. The iconoclasts, basing themselves on a tradition that claimed to be Oriental and placed the divine above everything else, including matter, simply ignored science. Scientific expertise did not interest them because it was a material kind of knowledge that ought to be disdained in favor of true knowledge of the divine. The iconodules, in contrast, identified more closely with Helleno-Latin culture; they could not imagine the divine without material representation of it. For them, the material world created by God was too important to be overlooked. It deserved exploration.

The rise to power of iconoclast emperors severely undermined science education. According to two chronicles of the ninth century, the first iconoclast emperor, Leo III (r. 714–41), gave the order not only to close the imperial university but also to burn it down—building, library, and professors. Although Byzantinists regard this story an iconodule legend, it nevertheless reveals how the iconoclasts were viewed by their enemies.[14]

The ascent to the throne of the iconoclast emperors accelerated the slide of science into decadence that had begun under the reign of Justinian. Ironically, the debate over icons, which initially led Byzantine society to downgrade forms of knowledge such as science, after a few decades incited the warring theologians, especially the iconodules, to turn to secular learning to draw arguments against their adversaries. They began studying Aristotelian logic anew, prompting an intellectual revival in the ninth century called Byzantine humanism.

John of Damascus, a Scholar of the Iconoclast Period

John of Damascus (c. 676–749) and Theodore the Stoudite (759–826) are the best-known scholars of the iconodule movement. Both studied secular science, but it seems that Theodore concentrated on the *trivium* rather than on mathematics. John, one of the most important Orthodox theologians of the eighth century, served as chief administrator to the Muslim caliph of Damascus. In the mid-730s he retired to the monastery of Saint Sabas and received ordination as a priest. Because John supported icons, the iconoclast Leo III tried to have him condemned by the caliph. John wrote important doctrinal treatises and was one of the first Christian authors to engage in polemics with Islam. Unlike Theodore, he had studied the *quadrivium* and was interested in the sciences. This great theologian left us one of the very rare texts of the eighth century that addresses science. In his *De fide orthodoxa*, John discussed not only cosmology but also practical astronomy, giving elements of astronomy such as the dates of the entry of the sun into each of the signs of the zodiac. This information was important because it determined not only the equinoxes (the spring equinox is the basis for the calculation of the date of Easter) but also the length of seasons. However, despite the fact that John was considered by the Byzantines a great savant and an expert in mathematics, he made gross errors in his calculations: he placed the spring equinox three days late and was mistaken by several days in calculating the length of the seasons.[15]

On the theoretical plane, John regarded theology as the queen and philosophy and the other sciences as her servants. According to him, the truth is revealed to us; it remains to humankind to find the means (through philosophy, including the sciences) to express it through reason. John of Damascus attacked superstition and ignorance; he called for the explanation of the phenomena of nature. In the lineage of the first fathers, he denied that the universe or its components had any sort of soul: "Do not complain about matter; it is not vile, nothing is vile that was created by God. It is the Manicheans who believe that."[16] Although he wrote the astronomical treatise mentioned above, John was not very interested in science. As we have seen, astronomers were adepts of the cosmology of Ptolemy and thus followed the cosmology of Saint Basil and not that of the school of Antioch. Yet John declared he was not interested in knowing whether heaven was spherical or vaulted; the only thing that mattered was that everything had been made according to the will of God.[17] Accordingly, he did not take a clear position on the debate over the spherical or conical form of the earth. He thought that reason is expressed only through faith. Despite this, in *De fide orthodoxa* he gave a good description of the world based on Hellenic cosmology and inspired by Basil's *Hexaemeron*, integrating into it some elements from the school of Antioch.

The world described in *De fide orthodoxa* is spherical, composed of the four elements in addition to heaven, ordered in space according to Aristotle's scheme.[18] But against Aristotelianism, John discussed the nature of air as an element, whether it might come from either fire (as residue of combustion) or water (the substance of its evaporation). The created world has a total duration of seven millennia (hence, its end is forecast for the year 1500). The spherical world is ordered like Basil's, but contrary to him and in accordance with the Antiochians, John affirmed that the angels inhabit the heaven of Creation and that the Kingdom of Heaven after the resurrection will be found in this same heaven and will last a thousand years. He gave the same explanation as Basil for the creation of day and night before the creation of the sun (cyclical movement of light), as well as for the formlessness of the earth (it did not have plants, etc.). Heaven is divided into seven zones, occupied by planets, and it is composed of very fine matter. The firmament moves from east to west, whereas the planets move from west to east. John also closely followed Basil's *Hexaemeron* as regards the luminosity of stars, which results from the light that God has furnished them: these bodies are merely the supporting matter for this light. As for the nature of light, he moved away from Basil (who spoke of divine light) to talk about "proto-created" light. The stars serve to measure time; the revolution of a heavenly body is more rapid the smaller the circle it follows: therefore the sun runs through a sign of the zodiac in a month, whereas the moon takes two and a half days. John innovated when he maintained that the water above the firmament serves to keep the latter from burning up from the heat of the fire of the stars.

John also advanced a theory of comets that does not agree at all with the Aristotelian idea that these bodies are phenomena of the atmosphere. According to him, comets are stars created to mark the death of a king, after which they disintegrate. In the case of Christ, the star seen by the magi was created only for his birth and evidently did not partake of heaven's movement. John, although accepting the idea of human free will, did not reject astrology entirely. In effect, the stars can be signs of events, but in that case it is up to the will of human beings to submit to them or not. He also gave an incongruous explanation of the incommensurability of lunar and solar calendars. When the moon was created along with the sun on the fourth day, it was full; hence, it was already almost 15 days old (counting from the new moon). By subtracting 4 days (because the moon was created full on the fourth day of Creation), the moon and the sun thus have a difference of 11 days. The twelve lunar months contain 354 days (a lunar cycle of 29.5 days according to John); the bissextile year contains 366 days, hence 11 more. In this way, the moon gives back to the sun the 11 days it owed it, for the 366 days depend on the solar cycle, whereas the 354 depend on the lunar cycle.

In early and middle years of the Middle Ages, in Byzantium as in the West, some monasteries housed libraries and copying centers, but, unlike in the West, Byzantine education was mainly provided elsewhere. The monastic movement linked to asceticism was born in the third century and was very important in Byzantine society. The monks, unlike those in the West, were not organized into religious orders, and each monastery had its own rules (*typikon*). Moreover, there was a strong individualistic tendency among monks, who organized themselves into small communities or became hermits. During the iconoclast period, as educational institutions declined, monasteries would be of greater importance for the preservation of secular knowledge.

A tough veteran of the movement and an advocate of monasticism, Theodore the Stoudite contributed to preserving the art of copying manuscripts in the famous scriptorium of his monastery, Saint John of Studios, where the technique of cursive writing was probably perfected. During Theodore's time, the Saint John of Studios Monastery was promoted as the main Byzantine scholarly center and also a center of resistance to iconoclasm. Theodore was a model of the eighth-century scholars who, though erudite, completely neglected scientific matters. In fact, his center did not produce any scientific manuscripts.

The rarity of intellectuals such as John of Damascus, one of the few who showed an interest in science during the iconoclast period, supports the claim that scientific expertise in this era had sunk to its lowest level since the advent of Christianity. The lack of interest in theoretical science was compounded by practical ignorance. A mediocre scientist, John asserted himself not only in the theological domain but also in politics. In *De fide orthodoxa*, to justify the submission of people to secular power, he used the allegory of the stars: the moon has a light borrowed from the sun because God wanted to show that there is a hierarchy in the world, that there exist a lord and his subjects. Therefore, one must submit to God but also to those who have power on earth by his grace, and one should not ask questions about where this power comes from, but accept it, thanking God.[19]

Astrology, Orthodoxy, and Islam

The iconoclast period saw the first scientific contacts between Eastern Orthodoxy and Islam. The dramatic expansion of Islam after the mid-seventh century took place in large part at the expense of the Byzantine Empire. Muslim Arabs conquered Alexandria, the empire's scientific capital, in 646 and turned its school into a Muslim institution. The Byzantine Empire was in constant war against the Muslims, and the whole population felt the Islamic threat. Any such period when the future appears uncertain is favorable to astrology, and now people hoped it might be able to predict the political future of the empire. Although all the fathers of the church, whether they belonged to the schools of Alexandria or Antioch, had condemned astrology as regards its capacity to predict human actions, the very orthodox Byzantines soon indulged in this art, circumventing the fathers' condemnation of the practice. The imperial court needed astrologers in order to make decisions; so it, too, paid little attention to criticism from the church. A practical and useful art, astrology thrived during the iconoclastic period, which meant that knowledge of mathematical astronomy also survived, because it was absolutely necessary to this art.

On the Arab side, astrology was truly in vogue. The Persian Abū Ma'shar (787–886), one of the most famous philosophers and mathematicians of Islam, was also a well-known astrologer. During the period discussed here, the Arabs were very open to the influence of Byzantine astrologers because they lacked sufficiently skilled practitioners. The caliphs welcomed them to their court. Later, the Byzantines invited Arab astrologers to their court.

Beginning with Constantine's reign, the Orthodox Church had become part of the state apparatus. Although intransigent concerning dogma and passionate about religious debates such as the one over icons, the church let the emperor do what he wanted when it came to astrology, even though the fathers' interdiction was clear. This condemnation had been based on the doctrine of free will, which alone could bring a person to sin or virtue, as well as the condemnation of the pagan idea that stars were divinities. The fatalism implied by astrological predictions was unacceptable to Orthodoxy because it would annul humankind's constant struggle to lead a life according to God's prescriptions. The fathers had said that God had created the stars to serve as signs to men, both to tell time and to forecast natural phenomena, but stars could not have a hold over human destiny, for in that case their nature would be divine.

Back in the fourth century, Basil had thought it completely valid to base oneself on the observation of "luminaries" (the sun and moon) in order to predict natural phenomena such as the passage of time. Moreover, Basil had said, the Lord has told us that the signs of the end of the world would appear in the sun, the moon, and the stars. He added that "some exceed the limits and draw from Scripture to defend the horoscopic art; they say that our life depends on movement in the heavens, and that the Chaldeans could find in the stars indications of what will happen to us. They simply took the word of Scripture 'May they serve as signs' to mean that it does not apply to atmospheric changes nor to the passage of time, but to the fate that befalls us."[20] Thus, Basil was condemning the horoscopic art that consists of determining the a person's birth sign in order to predict his destiny.

All fathers agreed with Basil's condemnation but the discussion continued about what kind of natural phenomena could indeed be predicted by the stars. God, after all, has created them as signs in the sky. So, can weather or natural catastrophes be forecast by the stars?

Gregory of Nyssa had written around 395 the treatise *Against Fate* especially to refute astrologers. *Against Fate* does not have the character of a theological work: Gregory's arguments were based on logic and observation. In the form of a conversation with a philosopher who is defending astrology, he dismissed each thesis proffered by the other, refusing to accept that the stars have any influence on the fate of humans, on society—and even on natural phenomena like catastrophes. For example, how does it happen that in a natural catastrophe various men perish who have had their individual horoscopes done and hence have life predictions that are so different from each other?[21]

Diodoros, bishop of Tarsus (d. 392) and one of the masters of the school of Antioch, had also firmly condemned astrology, for the prediction not only of human destinies but also of phenomena such as drought or rainfall: "If everything is governed by geniture [planetary configurations], why does a planet when it enters into a humid sign of the Zodiac with which it has affinities, not fill the earth with water in a single stroke, and why does one region overflow and why at the same time does another area lack rain, although the two are not very far from each other?"[22]

Despite this banishment of any science of prediction, astrology flourished in the very Orthodox Byzantine Empire, even during the iconoclast period. And despite Leo III's victory over the Arab fleet that was laying siege to Constantinople in 718, Byzantine society continued to feel endangered throughout the eighth century; infidels threatened at any time to invade the empire and to persecute Christianity. In this context, common people, eager for any signs of an uncertain future, ignored the condemnations of astrology by the church fathers.

One of the leading Orthodox astrologers of the eighth century was Stephen the philosopher (sometimes called Stephen the astrologer or Stephen of Alexandria), not to be confused with his homonym, a contemporary of Emperor Heraclius. This Stephen wrote three astrological texts that have survived: a treatise predicting the fate of Islam, an astronomical text *On the Mathematical Art and on the Peoples That Use It*, and an astrological text on the properties and relations of celestial bodies. Stephen was one of those rare scholars who knew astronomy and thus could draw from it what was necessary to make astrological predictions, in our case, to construct the *themation* of Islam, that is to say, the position of the stars at the moment of this religion's birth that would determine its future. To give weight to his predictions, Stephen indulged in a practice that had become common since the end of antiquity: signing his text with a false name, that of a known scholar. It sufficed to add "of Alexandria" to his own name to make people believe his predictions were the work of the famous savant who lived a century and a half before him, and consequently he dated his manuscript so that it could have come from his celebrated predecessor; thus, the text is falsely dated to the year 621. Not surprisingly, all the predictions concerning Islam proved correct up until 775, but were false thereafter—impressive evidence the text was written about 775.[23]

This dating, as well as the identification of the author as the writer of the second text (*On the Mathematical Art*), enables us to affirm that the Byzantines welcomed Arab astronomical expertise from the start, and especially during the Byzantine "dark ages" of the iconoclast period. The evidence is in the text: the author gave Constantinople a latitude that had been adopted by Arab astronomers, in the middle of the fifth *climate* (one of the seven ancient Greek zones of earth's latitude), and not given by Ptolemy (at the border between the fifth and sixth); he spoke of the *neōteroi* (new ones or moderns in relation to the old) who use the years of the Persian kings and Arab years; finally, he declared that the tables of Ptolemy included an error of five degrees for the sun and that the calendar in use was not practical. Given all that, Stephen, who had recently returned from a trip to Persia, implied that he adapted into Greek for use in Constantinople some tables of Persian or Arab origin. Stephen's reputation spread to the Muslim world. Abū Maʿshar cited him, as does an anonymous treatise of Caliph al-Ma'mūn's (r. 813–33) time; one of Stephen's books was cataloged in an Arab list of astrological books.[24]

Stephen defended astrology publicly and passionately. The complete title of his second book made his intentions very clear: *On mathematical art and on the peoples that use it: For those who say that it leads to sin. On the fact that someone who does not accept it commits an error. On its utility. And on the fact that it is the most precious of techniques.* Thus, Stephen was writing to refute the whole argument of the Orthodox fathers, and doing so in the midst of the iconoclast period when the Orthodox "fundamentalists" were in power. He declared in the first chapter: "I, coming from Persia and finding myself in this Happy City [Constantinople] and finding that the astronomical and astrological part of philosophy was extinct there, I thought it was necessary, my very dear and precious child Theodosius, to expound this doctrine in an easy manner and thus rekindle this science worthy of being loved, so that I would not be excluded, and that I would not be among those who hide their talent. This was neglected here because of the difficulty of expounding the tables and because calculating certain things is culpable."[25]

This statement tells us either that Stephen had perfected his practical astronomy (to understand and adapt astronomical tables and hence be able to calculate the position of stars at a given moment) in Persia, or that there no longer existed an able teacher in Byzantium at that time, or else that teaching this knowledge was not permitted. The statement also indicates that those who practiced astrology in Byzantium were hiding their talent, evidently because of the church's condemnation, and that Stephen's goal was to reestablish the splendor of astrology and make it a worthy science for society, in order that he himself would be recognized and could exercise his craft in tranquillity. It is very probably for this reason that he antedated his *themation* of Islam. If the Byzantines managed to be persuaded that the predictions for the fate of their enemy were correct, they might forget about the religious interdiction and adopt astrology—and astrologers.

Theophilos of Edessa (c. 695–785) was another prominent Orthodox astrologer. Leaving the Byzantine Empire, he entered the service of the Arabs and became at the end of his long life the chief astrologer of Caliph al-Mahdī (775–85) in Baghdad. He translated several Greek scientific works into Syriac. Theophilos predicted that Islamic domination would last 960 years—the duration of a grand stellar conjunction. The Arabs' welcoming of Theophilos as a great astronomer shows that astronomy had not gone extinct in Byzantium but rather was not openly taught (for a while) because of its astrological connections. Moreover, Theophilos's sources were largely Greek. Apart from a Sanskrit work of the sixth century, he based his work on the writings of Dorotheos of Sidon, Hephestion of Thebes, Julian of Laodicea, and Rhetorios—all authors of Greek astrology treatises.[26]

The efforts of Christian astrologers returning from Muslim countries to Byzantium around the end of the eighth century would rapidly bear fruit, and astrology would be reestablished in the Orthodox collective consciousness, although the church would never restore it officially. Emperor Constantine VI (r. 780–97) was defeated by the Bulgars in July 792, reputedly because of the unfortunate advice of his astrologer Pancratios, who had used the astrological work of Theophilos. This defeat and, five years later, one suffered by the Arabs would sound the demise of Emperor Constantine; his enemies would put out his eyes and shut him up in a convent. Despite these unhappy events, the Byzantine court would henceforth have its certified astrologers, whose religion—Christian or Muslim—did not matter.

Another factor favoring the development of astrology in the eighth century was an expectation that the end of the world was imminent. Since the end of the fifth century, this prophecy had recurred often in the Byzantine world. At the start of that century, the chroniclers Annianus and Panodoros had calculated that the creation of the world occurred at the spring equinox of the year 5491 or 5492 BCE. Therefore, the year 6000 (alluding to a multiple of the six days of Creation) seemed to be a propitious moment for the end of the world, which would thus occur around 509. But this date passed without major problems, and new dates were sporadically proposed. Meanwhile, the *Pascal Chronicle*, written around 630, pushed back by several years the beginning of the world to 25 March 5508 BCE.[27]

During the eighth century, the prophecy of an imminent end came back in force. The date 741 was proposed, and then the awful date was pushed back to around 897. A whole genre of predictions and calculations entered into these prophecies: that the emperor who defeated Islam would then be vanquished by the Antichrist, that he would be the last emperor, that the duration of the world was equal to a certain astronomical cycle. Apart from the fact that astronomic calculations were necessary for this kind of exercise, the very recourse to astrological science was seen by the supporters of this science as a solution to the grave problem of the apocalypse.[28]

Popular Science

We have already noted that the principal characteristic of Byzantine literature of the eighth century was the almost total disappearance of secular literature, which was replaced by religious works. But this does not mean that the theme of nature was no longer found among literary preoccupations; natural phenomena did not cease to evoke society's interest, and writers were always pondering their explanation. In the absence of texts of natural philosophy, these questions appeared in theological texts.

Questions about nature and natural phenomena were discussed during the iconoclast period in a naïve, unscholarly manner; one finds them dealt with in popular texts—hagiographic or miraculous—that are written in question-and-answer form. In this kind of book that aims to recount the lives of holy men and to describe their miracles, a young student poses questions to his teacher, a holy man, who elucidates everything the pupil does not know, whether moral or material, about such topics as the nature of thunder, rains, snow, and earthquakes. In this kind of teaching, the form of the universe is rather like the school of Antioch's, and manifestations of nature are acts of God. Especially in cases where natural phenomena cause some kind of catastrophe (earthquakes, droughts, floods), God justly provokes these phenomena in order to punish humans for precise sins. But alongside this view of phenomena and naïve conceptions of the universe, Basil's teaching is also present. In the very commonplace questions and responses attributed to Anastasios of Sinai (seventh century), the author insisted on the influence of the natural environment on humans and their society. Anastasios followed Basil in refuting astrologers' arguments about the influence of heavenly bodies on the character and destiny of humans; instead he put forward much more concrete phenomena, such as the climate and seasons.

The conception of the world, the form of the universe, the stars, matter, the manifestations of nature, life—all were perceived by ordinary Byzantines through these popular, moralizing, and simplistic texts called "lives of the saints." This was the popular science of the day, founded on miracles and on divine intervention, but didactic in intention. This knowledge had its roots in the school of Antioch's philosophy of nature from the fourth to the sixth century, and popular conceptions of the world would not change very much in the course of centuries to come. What would change is that around the end of the iconoclast period, learned literature on the philosophy of nature would make a resurgence and would shape the image of the world among the Byzantine aristocracy.

CHAPTER FOUR

The Return of Greek Science

The First Byzantine Humanism

Science and religion in Byzantium entered a new phase in the ninth century, characterized by a revival of Greek science. Two of the most visible faces of this revival were John the Grammarian and Leo the Mathematician. John was born sometime during the last quarter of the eighth century. A brilliant student, he became a professor, hence the nickname Grammatikos (grammarian). He then entered a monastic order and became *hēgoumenos* or abbot of the monastery of Sergios and Bakchos in Constantinople. At the start of the iconoclast Leo V's reign (813–20), he took the side of the enemies of the icons, and in 814 the emperor charged him with looking among the manuscripts of Constantinople for a copy of the acts of the iconoclast council of 754, because the original had been destroyed by the iconodules. Mission accomplished, John was considered a man of great erudition, who had the ability to persuade his opponents with well-chosen arguments. John's knowledgeable command of the manuscripts was a new factor at the end of the iconoclast period. One of the signs of this change was that two of the rare scientific codices we have date from this era: the *Vat. gr.* 1291, an illustrated manuscript that includes Ptolemy's *Handy Tables* and other astronomical and astrological texts; and the *Leidensis BPG* 78, which also includes Ptolemy's *Tables*.

Tutor of the future emperor Theophilus (r. 829–42), John became Patriarch John VII under his reign, but he was dethroned in 843 by iconodule regents, specifically the widow of Theophilus, Theodora (mother of Emperor Michael III, who was then age four), and Theoktistos, an influential eunuch (and effective ruler until his execution in 855). What is of interest to us is his visit as Theophilus's ambassador to the court of Caliph al-Ma'mūn in Baghdad, at a time when interest in Greek science was very lively there. Among the most notable developments was the creation of a library where Arabic translations of Greek science texts were kept. Coming back to Constantinople, John persuaded the emperor to reconstruct his palace in Vrya (in Bithynia) in the manner of al-Ma'mūn's and to add a number of automatons, which he helped fabricate.

Although the Byzantines seem to have abandoned the Greek tradition of making scientific instruments, they inherited the technology of automatons, developed by Heron of Alexandria (first century BCE), including the technique of gears, developed in a magnificent way by the first century BCE in devices such as the famous planetary called the "Antikythera mechanism." In spite of the fact that we have no mention that such mechanisms existed in the Byzantine era, devices like those constructed at Vrya to impress visitors are evidence of the perpetuation of a tradition of working with gears.[1]

After John the Grammarian's death, the iconodule patriarchs who succeeded him accused John of magic and had his remains disinterred and burned. His enemy and immediate successor, the patriarch Methodios, said of him that "he envied the lives of Pythagoras, Saturn, and Apollo," and the Byzantine chronicles described him as indulging in sordid practices in his dark laboratory.

Why were such accusations leveled against Patriarch John? Could it be related to his close involvement in science and technology? We know that astrology and alchemy belonged to the sciences in the eighth century, and any savant such as John who loved the sciences would be interested in them. If we add the crafting of automatons, this was sufficient evidence for his enemies to formulate accusations of magic at a time when the sciences were just beginning to be reestablished within the Byzantine Empire.[2]

At the end of the iconoclast era, interest in the sciences was resuscitating. We saw in the preceding chapter that one of the reasons for this renaissance was the renewal of interest in Aristotelian logic because it could furnish arguments to both parties, iconoclast and iconodule. Study of logic led inexorably to rereading philosophical texts and, hence, to natural philosophy. But whatever the root cause, very soon there was a veritable revival of the arts and sciences. This revival was called by the French Byzantinist Paul Lemerle (1903–89) the "first Byzantine humanism" because of the vivid interest shown by scholars of this period in ancient Greek literature.[3] Although the iconodule party was more open to secular knowledge, the two first figures contributing to this renaissance were the iconoclast patriarch John the Grammarian and his young cousin and protégé of an iconoclast emperor, Leo the Mathematician.

Leo was born around the turn of the ninth century. Son of an aristocratic family, he found that in the iconoclastic context the teaching of sciences in Constantinople left something to be desired; he could find a teacher for the *trivium* but not for science (the *quadrivium*). In the absence of not only educational structures but also a tradition of education, he sought a master with whom he could deepen his knowledge, eventually finding one on the island of Andros (near Athens, a week by ship from Constantinople). There he studied rhetoric, philosophy, and arithmetic. However, discovering that this teacher's knowledge was also limited, Leo ransacked the libraries of the monasteries of Andros for old manuscripts about science, finding a forgotten knowledge, still preciously guarded in the dark libraries of scattered monasteries. After years of research and reading, Leo recovered a whole domain that had been forgotten for more than a "dark" century during the debate over icons.[4]

The story handed down about the life of Leo claims that he followed a path of study that was characteristic of the Western Middle Ages: first find a teacher to acquire the basic knowledge and then dig into the monastic archives for manuscripts that were either hidden or forgotten. In Byzantium, this practice became a literary *topos* found in many "lives of saints," whose intent was to glorify the hero: by claiming that the knowledge of his teachers was not sufficient, the student went looking for—and found by himself—knowledge that did live up to his aspirations. But it is possible that during the restructuring following the iconoclast period, the renaissance of science did go through a period of "rediscovery" that exceeded the teaching by acknowledged masters and thus required research into neglected manuscripts. Leo symbolized this quest and became the very image of scientific renewal.

The rest of Leo's story is mixed with legend. Returning to Constantinople, he gave courses in philosophy, specifically the *quadrivium*. One of his students, secretary of a general, was taken hostage by the Arabs, who sent him to Baghdad as a slave. There, at the court of al-Ma'mūn, Leo's student distinguished himself for his expertise and impressed the caliph's surveyors by showing himself more able than they in demonstrating the theorems of Euclid. When the courtiers asked him if Byzantium had many savants like him, he responded in the affirmative, saying that he considered himself only a student and that his master lived in poverty in Constantinople. And so al-Ma'mūn liberated the young man and sent him to Constantinople with a letter of invitation to Leo, to whom he promised enormous riches if he agreed to come. Leo wisely showed the letter to the emperor (Theophilus), who finally recognized his talent and granted him a salary to give courses in public at the Church of the Forty Martyrs. Meanwhile al-Ma'mūn insisted he come to Baghdad and even offered him a large quantity of gold, plus a treaty of eternal peace to the emperor if he would let him leave. But Theophilus refused, not wanting to reveal to his enemies the knowledge that was the glory of the Greeks. To console Leo, he asked John the Grammarian, then patriarch of

Constantinople, to name him metropolitan of Thessalonica. So—and this part of the story historians can verify—we find Leo as metropolitan of Thessalonica from 840 to 843. But in line with the fate of his iconoclast protector, he was soon dismissed by the iconodule regents.[5]

During the reign of Theophilus, Leo conceived the famous optical telegraph to transmit signals from the eastern frontiers of the empire to the capital. The novelty of this system was that it was based on two synchronized clocks, one on the frontier and the other in the capital. The clocks were divided into twelve intervals of time, and every two hours corresponded to a precise message. A message sent at midday signified "invasion"; at two o'clock, "war"; and so on. The signals were supposed to arrive within an hour.[6]

The involvement of Leo with the iconoclast emperor and patriarch did not have a negative impact on his career as a scholar. His continued service, combined with the fact that the emperor gave him a salary and privileges, shows that the prestige of science had indeed risen around the middle of the ninth century. Leo's legend (containing an element of truth) is quite different from the legend that claims that a century earlier an emperor had burned the university and its professors. In the story of Leo and al-Ma'mūn, the sciences are exalted and considered to be a great treasure for Byzantium.

The change in the attitude of secular power (which generally went hand in hand with that of the patriarch, who was appointed by the emperor) toward science is also attested by the palace's interest in higher education at the start of the reign of young Michael III (842–57). Bardas, the brother of the queen mother Theodora, was regent between 856 and 866 and uncontested master before being assassinated. He was one of the rare Byzantines to bear the title of kaiser without being brother to an emperor. This important personage showed great favor toward secular knowledge. Probably inspired by Arab policy to foster the sciences (the legend concerning Leo the Mathematician can be read in this way), he founded a school where science, especially mathematics, was taught. The School of Magnaura in Constantinople was founded around 855, after the royal palace of the same name. Leo the Mathematician was placed at the head of the philosophy department (which showed it was considered the most important subject), and his student Theodore at the head of the geometry department; Theodegios led astronomy, and Komitas led grammar. The professors were paid by the state, and tuition was free.[7]

The public education given at the Church of Forty Martyrs undoubtedly inspired the new school. From its curriculum, it seems evident that it was specifically a scientific school in which theology was completely absent. The school became an important institution that outlived both Bardas (assassinated in 866) and Leo (who probably died during the 870s) and would constitute the principal scientific institution in Byzantium for two centuries. During the reign of Constantine VII Porphyrogenitus (913–59), the school remained divided into four departments, with philosophy still at the top, and at its head the *protospatharios* Constantine. At this time, a *protospatharios* was equivalent to a member of the Senate, which no longer had any power. Constantine would later be appointed *eparch*, which means prefect of Constantinople. For the first time in Byzantine history, a high government official was serving as professor of philosophy, an occurrence that would become common later as the aristocracy mingled with the scholarly caste.

During the reign of Constantine Porphyrogenitus, state power took an additional step toward the institutionalization of nonreligious education when a school administrator was appointed for each town. The panegyric addressed to an emperor by a chronicler in the eleventh century signifies this change in attitude toward science: "He restored the sciences, arithmetic, music, astronomy, geometry, solid geometry, and all of philosophy, which had been neglected and lost for a long time because of the ignorance of powerful men, and he sought and found the best teachers in each domain."[8]

Religious support for literature and science had actually made its appearance earlier. The ninth century witnessed the first patriarchate of Photius (858–67), one of the most erudite scholars of the Byzantine world, author of a famous library catalog, in which he summarized the 279 books he had read (158 religious books and 121 secular ones). This is a fabulous and precious work, since almost half of the books he covered are lost today. A scholarly patriarch, lover of Greek literature, Photius observed with a benevolent (if not enthusiastic) eye the scientific renaissance that was taking place.[9]

It seems that Byzantine power, both political and religious, was ready to accept the cultural change brought about by humanism. Note that the patriarch was a political figure, often appointed by the emperor, not by the synod, and sometimes he had little relation with the clergy. Photius came from an aristocratic family and was a layman who held the offices of *protospatharios* and *protoasikritis* (the latter being equivalent to head of the imperial secretariat). In order for the synod to be able to name him patriarch, his protector Bardas arranged for him to enter holy orders and rise through the ecclesiastical grades—in only six days! Photius was criticized by clerics, who saw him as a mere layperson. Bishop Nikitas David depicted him as a true scholar but one interested only in secular knowledge. His vast expertise led him to be arrogant; instead of being a humble servant of God, he built on the rotten foundations of profane science.[10] Among society as a whole and the lower clergy, a legend existed that Photius in his youth had sold his soul to a Jewish magician.[11] It is true that there would never be a condemnation of Photius in the way the other "magus," the iconoclast John the Grammarian, had been condemned, but the sciences were still viewed with some distrust by Byzantine society.

A statement written by Constantine the Sicilian, a former student of Leo's, sheds additional light on attitudes toward his master. In a mock obituary, Constantine placed an anathema on Leo, who taught all the profane wisdom of which the ancients were proud but lost his soul in that sea of impiety. According to Constantine, Christ punished Leo as a renegade who venerated Zeus and then sent him down to Hades to join Chrysippos, Socrates, Proclus, Plato, Aristotle, and Epicurus, as well as Leo's favorites, Euclid and Ptolemy. Educated Byzantines were scandalized to see such ingratitude from a student to his master, but Constantine escaped punishment.[12]

Historians of science have often stressed the role of ninth-century Byzantium in safeguarding Greek scientific literature. In effect, the most ancient copies of most of the Greek scientific works that have come down to us do date from this era, as attested by the study of their offshoots found in later manuscripts. This was happening three centuries before the Latin West rediscovered the basic texts of ancient science and finally became interested in scientific knowledge. However, tension between Eastern and Western churches, together with the ignorance in the Latin West of Greek, the language of ancient science, prevented the spread of the scientific knowledge that had been safeguarded by the Byzantines to the network of the monasteries in Western medieval Europe. In contrast, the Arabs, who were already exploiting ancient Greek science, showed a vivid interest in Greek scientific literature as preserved and edited by Byzantine scholars. But while the Arabs tried to develop ancient Greek science by correcting Ptolemy and contributing to mathematics, the Byzantines treated Greek scientific texts in the same manner as the texts of the fathers of the church. They copied them, taught them, but rarely developed them.

As part of their revival of interest in the sciences, Byzantines copied ancient manuscripts, even when there were no specialists to exploit them fully. The case of astronomy is clear: the only new contributions that are extant are a table of thirty bright stars dated to 854, an updating of Ptolemy's table using the precession value of the great Alexandrian astronomer (one degree every one hundred years, which means that the stars ought to have moved seven degrees since

Ptolemy), and some elementary commentaries on his *Tables*.[13] What a contrast there is between this lack of expertise and the magnificent astronomical manuscripts, composed in uncials (despite the appearance of cursive writing) and sometimes illuminated, that have survived from this century. It is significant that four manuscripts of the *Handy Tables* are extant, the two mentioned at the start of this chapter, copied under the reign of Leo V (813–20), plus two others copied under the reign of Leo VI the Wise (886–912). The choice is not by chance, for the *Handy Tables* are (as their name indicates) a simplified version of Ptolemy's *Almagest*, laying out the practical side of his astronomy; because the tables were used to determine the position of stars at a given moment, they were highly prized by astrologers.

The contribution of Leo the Mathematician and his entourage (especially his students) to safeguarding science manuscripts was crucial in the ninth century. This group recopied all the extant works of Euclid, Archimedes, Ptolemy, and Proclus.[14] Around the middle of the century, Byzantines collected the works of Aristotle, followed by those of Plato.[15] Leo had not only rediscovered and taught Greek scientific texts but had copied them himself and had recruited copyists. Leo and his students considered themselves the heirs of Greek science and applied themselves to the task of saving dilapidated manuscripts before the sources of this knowledge disappeared forever. This represented a major change from the spirit that had prevailed only a few decades earlier, when the Byzantines judged ancient Greek knowledge to be pagan, useless, and dangerous for the health of Orthodox minds.

Hellenism and Orthodoxy: The Cases of Psellos and John the Italian

I have stressed the prime role played by the Byzantine emperors in relations between science and Orthodoxy. When an emperor was favorably disposed to secular knowledge, then teaching was encouraged and subsidized by the state, which gave science an important status that the clergy and society could not ignore. The history of Byzantine science is marked as much by the personalities of emperors as by those of scholars. Leo VI the Wise was himself a scholar who had studied under Photius. His successor, Constantine VII Porphyrogenitus, also strongly supported science. In contrast, Basil II the Bulgaroctonos (r. 976–1025), as his name Bulgarslayer indicates, was more concerned with killing Bulgars than with cultivating knowledge. Under this emperor, the institutions of higher education were once again abandoned. Two decades later, Constantine IX Monomachos (1042–55) took an interest in sciences and education and was assisted by his adviser Michael Psellos.

Michael Psellos (1018–78 or 1096) is the Byzantine scholar best known in western Europe. His name at birth was Constantine. Son of a civil servant, he studied in Constantinople and Athens and followed a political career. He took part in several intrigues that made or unmade emperors, which led to his being called the *paradynasteuon* (prime counselor of the emperor). In 1045, under the reign of Constantine Monomachos, as *protoasikritis* he also became *hypatos* (consul) of philosophers, the equivalent of rector of the University of Constantinople. Psellos seems to have been the principal instigator in reforms undertaken by Constantine to improve studies in Constantinople. In effect, in 1047 the emperor founded the faculty of law and named Psellos director of the philosophy faculty. It is not clear if Constantine founded a new institution or simply revamped what already existed by creating the post of *hypatos* of philosophers.

But in 1054 Psellos left the university in disgrace (for unknown reasons) and became a monk under the name Michael. He returned to Constantinople two years later when Empress Theodora Porphyrogenitus during her short reign (1055–56) recalled him to the capital. He again took an active part in court intrigues that shifted power until the accession of Michael VII Doukas (1071–78). Although Michael had been one of his students, he did not give Psellos a post as elevated as he desired. He was probably exiled to a monastery under Nikephoros III (1078–81).

Psellos's involvement in politics and his aspiration to the highest posts conferred on him an authority in science that went beyond that of a simple savant. Called the "universal man," he was also one of the most prolific writers of Byzantium; his writings deal with history, philosophy, theology, and science. His renown went beyond the borders of the empire, and he claimed that among his students were Celts, Arabs, Persians, Egyptians, and Ethiopians. Although he was not modest by nature, this assertion seems true.[16]

Psellos illustrates the complicated and ambiguous relations of the Orthodox world of his day with the sciences. On the one hand, he affirmed himself as an Orthodox believer who found in faith the answers to his spiritual questions. On the other hand, his curiosity and erudition in the secular knowledge of ancient Greece remained unquenched, and he went on to practice astrology, alchemy, and even magic. A Platonist, he knew and commented on Aristotle and admired the ancient Egyptians and Chaldeans. He boasted of his interest in five different civilizations: Chaldean, Egyptian, Greek, Hebraic, and Christian.[17]

Such pride at knowing the literature of non-Christian civilizations, as well as Psellos's practice of sciences condemned by the church (astrology and magic), gave his many enemies ample opportunity to attack him. An astute politician, he managed several times to survive these attacks by periodically making an act of loyalty to Orthodoxy, and he wrote a confession of faith during the reign of Constantine Monomachos. When he was accused of being under the decisive influence of Plato, he defended himself by maintaining that many elements of secular science were useful, citing Saint Basil and Gregory of Nanzianzus as corroboration.

To defend himself from accusations that he practiced astrology, he wrote a short declaration inspired by the church fathers that asserted that astrology is in contradiction with providence and free will.[18] And as for magic, he took cover behind his various acts of faith. But he was rather compromised, even publicly, when he wrote: "I will not reveal to you how to fabricate amulets that chase away sickness, for it is possible you will not imitate me correctly."[19]

If Psellos's acts of faith appear to be merely the maneuvers of a politician, there is in fact a text in which he seems to give an explanation from the bottom of his heart for his intellectual contradictions. In the funeral oration he gave for his mother, after saying that Christian faith cannot give an answer to every question, he affirmed that "because one cannot think that the life that was granted me suffices in itself, but that it is at the service of others and is going to be absorbed as from an overflowing vessel, for this reason I concern myself with idolatrous culture, not only in its theoretical aspect but also in its history and poetry."[20] Psellos was concerned with the knowledge of "idolaters" in order to communicate it; hence, this knowledge was worth being known and studied.

Psellos was a child of Byzantine humanism but at the same time a scholar in the Middle Ages. His love of the apocryphal sciences, to seek knowledge in supposed hidden meanings in texts and various symbols, led him to investigate Egyptian science, which, according to him, was the source of later science, including that of the Greeks. The discussion that bears on the roots of Hellenic sciences and especially the role of Egypt was not new, dating from at least the era of the historian Herodotus. Ancient Egypt had fascinated the Greeks and continued to fascinate Byzantine scholars. Psellos claimed that Pythagoras was the first to introduce Egyptian civilization into Greece and that Plato was wrong to believe that the Greeks had improved on the ideas of foreign peoples—on the contrary, the Greeks were lazy about searching for the truth, notably concerning God. Psellos even thought that Diophantus had been influenced by Egyptian methods of calculation.[21]

With respect to nonapocryphal sciences, Psellos's contributions were to mathematics, astronomy, and the philosophy of nature.[22] His best-known didactic work is *General Education,* which presents notions of natural philosophy.[23] However, his unequaled renown in the sciences caused a number of scientific texts to be misattributed to him, such as the oldest Byzantine *quadrivium* that has come down to us, entitled *Synoptic Treatise in Four Lessons,* which was written in 1008 (before his birth), as well as *On Natural Things,* actually by Simeon Seth.

Psellos comes across as an independent spirit who dared on several occasions to flout the commandments of the Orthodox Church in the interests of his curiosity, which ran from the traditional secular sciences to any sort of apocryphal science. If he was not reprimanded by the church, it was because he knew how to maneuver and, during the Doukas dynasty, had powerful support.

Nevertheless, Psellos's work was by no means subversive. Byzantines would positively remember his commentaries on the philosophy of nature and his texts of a practical nature, such as the calculation of the date of Easter or the catalog of minerals and their characteristics.[24] His name became a reference for Byzantine science in the following centuries, and his didactic texts on natural philosophy would be replaced only by those of Nicephorus Blemmydes two centuries later.

Psellos may not have had major difficulties with the church, but the same was not true for his student and successor in the post of *hypatos* of the philosophers, John Italos (c. 1025–90). As his name suggests, John was born in the south of Italy; his father was a Norman mercenary. Protected by the Doukas family, he settled in Constantinople around 1049, where he followed Psellos's courses, but he was soon arguing with him. Under the reign of Michael Doukas, he was accused of impiety for the first time, but the affair did not have any repercussions. John's most heretical ideas related to the incorruptibility of the world and to challenging the Neoplatonic thesis of the creation of the world. He argued that science alone could approach the truth, that ideas (like matter) were eternal, that miracles must have a physical explanation, and that there had been no creation *ex nihilo*—in short, things that would enrage even the most moderate theologians.

Around 1076, John was indicted by theologians for his impious theses about the creation of the world, but the case did not go to trial. A few years later, Alexius I Komnenos became emperor (r. 1081–1118). Though interested in science, Alexius was also a pious man who sought to re-Christianize higher education by introducing the study of Holy Scripture at the highest levels. Thus, for the first time, the patriarch was granted the right to supervise the content of higher education. It was not the most propitious moment for John Italos, who had neither the breadth of knowledge nor the diplomatic capabilities of Psellos. He lost the emperor's favor, was again accused of impiety, in 1082, and this time went on trial. The court condemned him as a heretic and a pagan for his philosophical ideas, such as his denial of the creation *ex nihilo*, which came into flagrant contradiction with dogma. The court sentenced him to perpetual reclusion in a monastery—but only after a sequence of eleven anathemas were pronounced against him.[25]

The condemnation of John Italos provided an opportunity for the Orthodox Church to condemn secular study more generally. Taking advantage of the occasion, the synod added to the Sunday service the following reading: "Upon those who indulge in Hellenic studies and do not study them solely for education but follow their futile opinions—anathema." This official condemnation of some ideas associated with Hellenic science (e.g., the eternity of the world) has been read in Orthodox churches up until the present day. Note that it does not forbid the study of Hellenic sciences—provided that they are considered not as true but only as part of a general education, as exercises of the mind.

An Eleventh-Century Manual of Natural Philosophy: Seth's Physics

The reforms by Alexius I did not signify the abandonment of secular science, which very probably continued to be taught at the same level as before, but henceforth theology would crown these studies.[26] Ironically, the very Christian Alexius, according to the testimony of his daughter, kept at his court four astrologers, two of whom were Egyptians, one an Athenian, and the other Simeon Seth.[27]

Simeon Seth, a contemporary of Psellos's, was an astrologer and doctor who at one time probably withdrew to a monastery in Bithynia, a common practice for Byzantine dignitaries who had fallen into disgrace. Later he appeared in Egypt (1057–59) and at the courts of Michael Doukas and Alexius I Komnenos, where he must have ended his days.[28] Emperor Alexius's daughter, Anna, described him as an able astrologer who predicted the date of the death of Robert Guiscard, a Norman warrior well known to the Byzantines because he had contributed to the Norman conquest of Italy. Seth knew Arabic so well that he had translated some animal fables from Arabic. His best-known works were *On Natural Things*, a treatise in five books; *On the Properties of Foods*, in which he described 228 sorts of plants and animals and advocated Oriental medications rather than those of Galen; and *On Beer*.[29] *On Natural Things* circulated widely in the Byzantine and post-Byzantine world. No fewer than twenty manuscripts of it have come down to us, dating from the thirteenth to the eighteenth century. It was even translated into modern Greek at the start of the eighteenth century. This treatise, dedicated to the emperor Michael Doukas, presented the philosophy of nature as popular science. Seth showed that the earth is spherical by using traditional arguments, such as the difference in the local time of observations of eclipses, the visibility of parts of the sky that depend on latitude, and the appearance of mountaintops before low-lying land when ships approach a coast. He explained the division of latitudes of the northern hemisphere into seven climates and gave the size of the *oikoumenē* (the inhabited earth that extended from China to the Canary Islands) as twelve hours, which was equivalent to half the earth's circumference. He attributed phenomena such as rain, hail, snow, thunder, lightning, and earthquakes to natural forces. Indeed, Seth maintained that everything in the sublunary world had a physical explanation and could be explained by the characteristics of the four elements interacting with the heat or light from the sun. God never intervenes.

Seth invoked God only in discussing the plurality of worlds, where he mentioned that certain philosophers believed in the existence of multiple worlds, each with its own human beings and animals. He repudiated the notion, held by certain Greeks, that the heavenly spheres had souls; for him, all celestial movements were purely natural. Concerning the important theological question of the incorruptibility of the world, he presented Aristotle's belief that the world is uncreated and that heaven is incorruptible and Plato's claim that it is created and incorruptible, and then he advanced its own thesis that all bodies have a limited force that is renewed by diurnal movement. The world is made of order and disorder; things here below are irregular and without order, in the sky irregular and ordered, and in the beyond regular and ordered. Seth defended the Platonists against the thesis of the Aristotelian fifth element by borrowing his arguments from Philoponus. He also followed Ptolemy's cosmology based on epicycles and not Aristotle's concentric world system. But he referred to Aristotle in declaring that a void cannot exist either in the created world or outside it; for the void is a place without bodies that can receive a body, and in the outside heaven there is no "place" at all but instead the spiritual world. Thus, except for a few passing references, one could scarcely distinguish between Seth's text and a work of pagan Greek philosophy. Nevertheless, the church left him undisturbed. No doubt his astrological abilities led Alexius I to pardon his deviations from Orthodox teaching.[30]

Astrology flourished during the Komnenos dynasty (1081–1185). The emperor Manuel I (1143–80) himself practiced astrology so much that the court grammarian, Michael Glykas, dared to attack the imperial ruler, using arguments dating back to the church fathers. Manuel defended himself against Glykas's accusations by addressing a letter to an anonymous monk. Glykas, far from submitting, replied with a book that explained the difference between astronomy and astrology to demonstrate the utility of the former and the impiety of the latter.[31] Glykas was somewhat exceptional in his opposition to astrology. His contemporary John Kamateros, the best-known astronomer of his day, took a great interest in the practice. He dedicated two astrological poems to Manuel and wrote a treatise on the astrolabe, the primary instrument used by astrologers. It is notable that the arguments advanced by enemies of astrology remained the same as those of the church fathers. Nine centuries after Basil the issue returned, and this time it was the Christian emperors, not the pagan philosophers, who were proving to be "impious."[32]

Despite the varying attitudes of emperors, the sciences in the twelfth century were well anchored in both higher education and the mentalities of Byzantine savants. New schools of higher education appeared, and the one in Thessalonica achieved sufficient importance to attract students from the capital. The Orthodox Church became reconciled to the fact that secular science should be part of the curriculum of educated people; extreme reactions against scientific education (as in the case of John Italos) occurred only if someone flagrantly contradicted what had been taught by fathers such as Basil or Gregory of Nyssa.

We can get an idea of what constituted secular science for an educated Byzantine around the second quarter of the twelfth century from a letter that Michael Italikos addressed to the empress Irene Doukas: "I know these sciences [geometry, arithmetic, and agriculture] and do not deny them, and I have checked Aristotle and Plato and verified the periods of stars and their many constellations, and I swear on the sacred head that I do not overlook what Hipparchus and the very scholarly Ptolemy have said about astronomy or what the very mathematical Aristarchus has written."[33] Basic science education no doubt consisted of the *quadrivium* plus the Aristotelian philosophy of nature. Those seeking more advanced knowledge would have to study Euclid, Diophantus, and Archimedes in mathematics; Theon and Ptolemy in astronomy; and Galen and Hippocrates in medicine.

We have said little about medicine so far, but the Byzantines excelled in this area. Building on a long-standing tradition of charitable enterprises, they began in the fourth century to develop hospitals, not only to provide food and shelter for the sick but also to treat their illnesses. Apparently for the first time in history, the Orthodox staffed their institutions with physicians, pharmacists, and medical assistants. They also used these hospitals to train medical workers. Today the best known of these hospitals is the Pantocrator Xenon, created in 1136 by Emperor John II Komnenos as part of a monastery in Constantinople. According to the hospital's surviving rule book (the *Typikon* of the monastery), there were fifty beds divided into five sections or wards: for patients suffering from wounds or fractures, for those with diseases of the eyes or intestines, for women, and two for men. Each section was attended by two physicians, three ordained medical assistants, two additional assistants, and two servants. Female physicians and staff ministered to the women patients. In addition, the hospital employed five pharmacists and operated a separate infirmary for the monks and an outpatient clinic. It stood in the community as a monument to Christ the healer.[34]

The institutionalization of scientific study, the privileges given to professors, and the involvement of the emperor and the patriarchate all gave status to science, which became part of the Byzantine culture of the eleventh and twelfth centuries. By the end of the twelfth century, the Patriarchal School in Constantinople seems to have become much more important than the one financed by the emperor; apart from theology, students learned rhetoric, medicine, philosophy, and mathematics. The school created a new chair, *maistor* (master) of philosophers, second in prestige only to the *hypatos* of philosophers, named by the emperor. But the centralized system of Byzantium, together with the importance of the town of Constantinople compared to the other towns of the empire, prevented the development of the high schools newly created in other cities. The school of Thessalonica, the second most important city of the empire, was an exception, but it soon declined. During the same period, in the Latin West the newborn universities multiplied. These universities, unlike Constantinople's high schools, had a certain autonomy, at least in the appointment of professors. In centralized Byzantium, the emperor and the patriarch nominated their protégés as heads of the university and the Patriarchal School. Thus, the status and the protection provided to science by the heads of the empire carried the seeds of stagnation. Although scientific teaching progressed, there were no vigorous discussions of scientific matters, and Byzantine contributions to science remained marginal.

CHAPTER FIVE

Struggle for Heritage

Science in Nicaea and the Byzantine Renaissance

On 12 April 1204, Christian crusaders from western Europe, thwarted in their effort to recapture Jerusalem from the Muslims, instead attacked their fellow religionists in the East and sacked the city of Constantinople. Thousands of combatants of the Cross, dazzled by the riches of the most splendid city of Europe, threw themselves on its treasures. Needless to say, this attack during the so-called Fourth Crusade permanently damaged relations between the Eastern Orthodox Church and the Roman Catholic Church. The Byzantine Empire, weakened during the previous century by wars and interminable struggles for power, could not resist the attacks of the Crusaders, who moreover had been summoned by Alexius IV, the son of the fallen emperor Isaac II, to help him against his adversaries. After his coronation, Alexius, sensing the danger from his ephemeral allies, tried to remove them as quickly as possible from the city, but this proved a futile effort. The wealth of Constantinople was too overwhelming in the eyes of soldiers, who had seen nothing comparable in the course of their travels across medieval Europe.[1]

Comparative price calculations show that at this time a manuscript of two hundred pages could be worth a two-room dwelling in Constantinople. The soldiers of the Crusader army might not be able to read, but they must have been aware of the value of these beautiful parchment books. Therefore, they carried Byzantine manuscripts back to the West, where they would later constitute the kernel of several European libraries, including the Vatican's. They would be fully exploited by Western scholars only two centuries later, when the Greek language was again taught after the arrival of a second wave of Greek manuscripts, the result of another conquest of Constantinople, this time by the Ottoman Turks.

The conquest by the Crusaders completely changed the institutional landscape of Byzantium, including that of science. The emperor and the patriarch fled. Schools of higher education, which had been state institutions financed either by the patriarchate (the Patriarchal School) or by the emperor (the University in its various forms), immediately ceased to exist. If we add the destruction of libraries and the disappearance of manuscripts, the catastrophe seems to have been absolute. Yet this catastrophe brought about an unexpected renaissance.

Nicaea: A Medieval State That Loved Science

At one hundred kilometers from Constantinople lay the town of Nicaea. There, Theodore I Laskaris (r. 1204–22) established his court and in 1205 founded the Empire of Nicaea, which at the start included only this one city and its adjacent provinces. But the symbolism was heavy with meaning. By calling this little country an "empire," Laskaris was claiming it to be the successor to Byzantium, in the face of two other Greek states that had just been created, the Despotate of Epirus and the (minuscule) Empire of Trebizond.

Theodore I's aim could not be achieved unless Nicaea asserted itself as a worthy cultural successor of Constantinople, and for that it had to establish mechanisms for the transmission of knowledge. More than a symbolic matter, this need for educational structures of a superior level became pressing when it came to finding civil servants to serve this new state, for Nicaea had nothing: neither libraries, school buildings, nor great scholars. The scholars would soon arrive, fleeing Latin-dominated Constantinople. Scholars, who simultaneously occupied key administrative posts, established themselves at the court of Laskaris, who gave them the task of organizing education in the new empire. For this purpose, but also in order to assert that the heritage of the Byzantine Empire was now found in Nicaea, Theodore I regenerated the post of *hypatos* of philosophers, which he granted to Theodore Eirinikos, a former high-ranking officer at the court of Constantinople. The fact that Eirinikos would have a lofty seal that bore his title of *hypatos* shows the concern to legitimate educational authorities even before they could really function.

The post of *hypatos* of philosophers was important in the Byzantine state hierarchy; it occupied the twenty-ninth rank among imperial offices. It was sometimes granted to scholars who already held high governmental responsibilities, which implied an interaction between the government and higher education. Sometimes, this post was a prime spot for advancement in the governmental (or even religious) hierarchy. Theodore Eirinikos passed from the *hypatos* of philosophers to become patriarch in 1214, restoring the tradition of philosopher-patriarchs. At Nicaea, the demarcation between secular science and the upper hierarchy of the Orthodox Church seems to have blurred. The successor to Eirinikos, Demetrios Karykes, was appointed by the emperor as spokesman for the Orthodox Church in negotiations with the papal legates in 1234 over questions of dogma that separated the two Christian churches. However, his failure to promote Orthodox views in the face of the Roman Catholics, who were better prepared for the debate, did not have negative consequences for the secular science that he also represented. On the contrary, the emperor interpreted this failure as the product of a lack of secular educational structures; and though he replaced Karykes as *hypatos* of philosophers, he redoubled his efforts to promote higher education.

The project of Theodore I to revive Byzantine secular knowledge in Nicaea had unforeseen consequences. In this tiny empire without schools and without the wealth of Constantinople's libraries, the sciences would soon occupy a very important place in the state ideology. To be educated and to have access to scientific knowledge were important, and the emperor and patriarch were both advocates of transmitting this knowledge to train future state officials. The best-known scholar of this period, Nicephorus Blemmydes, taught science in his capacity as *logothetēs* (chancellor) of the Great Church, under the aegis of Patriarch Germanos II. In 1240 Blemmydes became tutor to the young crown prince, Theodore, whom he initiated into the sciences. For six years, the future emperor Theodore II (r. 1254–58) studied philosophy, the *quadrivium* (arithmetic, geometry, music, astronomy), and physics. Then he continued his studies with Akropolites, himself a student of Blemmydes', who had become a high official of the Nicene state.[2]

The preponderant place occupied by secular knowledge at the Nicene court would have a major influence on the cultural identity of Byzantium. The word *Hellenic* (i.e., Greek), which had had negative connotations in Christian Byzantium, acquired under Laskaris a positive significance. The emperor regarded his people as descendants of the ancient Hellenes, his army as the Hellenic army, and Asia Minor as the Hellades. Culminating this return to the Hellenic ideal, Theodore compared Nicaea to Athens during its golden age, even maintaining that it surpassed it, since Nicaea possessed both secular and Christian philosophers. Theodore's use of the term *philosophers* to designate theologians must have been shocking for the Orthodox, who made a clear separation between philosophy as a secular and often profane science and theology.[3]

The Nicene people were far from convinced by Theodore's aspirations, but the idea of Hellenic identity germinated among Byzantine intellectuals. Despite the Nicene state's conquests, it never compared in extent to the vast Byzantine Empire. At its largest expansion, Nicaea controlled only a large part of Asia Minor,

Thrace, and Macedonia, encircling Constantinople, which remained under Latin domination. Instead, it sought an identity in intellectual achievement.

Theodore's dreams remained far from reality. Despite all his efforts, the most elementary scientific knowledge remained beyond the scope of the empire's officials. Even his mother, Empress Irene Laskaris, grand protectress of the sciences, as well as the *aktouarios* (court physician) of the empire, appeared not to know the explanation of solar eclipses when she attempted to refute the correct explanation supplied by the young George Akropolites, who had just studied astronomy under Blemmydes. Akropolites, despite understanding the mechanism of eclipses, later maintained that the eclipse of 1239 was a sign of Irene's imminent death.

We have seen that in Byzantine history there were periods when the sciences, and precisely the ancient Greek sciences, were well accepted as worthy knowledge by the Orthodox Church. But never until the Nicene period had the milieux of savants and the upper hierarchy of the clergy been so blended. In Nicaea, scholars made careers within the church and vice versa. When in 1234 Emperor Vatatzes (r. 1222–54), who had been a military commander, sent five young Nicenes to study with the famous professor Theodore Hexapterygos, he addressed his favorite, the young Akropolites:

> I send these pupils to the school, taking them from Nicaea, but I send you to be taught with them, taking you from my palace; prove that you really come from my house and exert yourself in your lessons. Had you become a soldier you would have received a salary from me, as much or a little more will you receive [for studying] since you come from a noble family; and if you become master of philosophy you will receive great honors and rewards, because only the Emperor and the Philosopher are the most famous of men.[4]

In this speech, the emperor placed philosophy above religion; he enumerated the most celebrated men—but did not mention the patriarch. This was a new kind of discourse for the Orthodox world, one that promoted the value of education and science but not the salvation of souls. The speech reflected the school of Christian Orthodox thought that wanted humankind to perfect itself by studying the nature created by God. For the first time, the Byzantines "had to consider that change was now an ever-present element in their lives, whereas the lives of their ancestors must have appeared to them considerably calmer and more anchored in the serene havens of a tradition that the passage of centuries had legitimated even more."[5]

Both culturally and religiously, Byzantines were very attached to tradition.

Change and innovation were not to their taste, and this was visible in science. Since the end of antiquity, neither the education curriculum nor physical theories had budged. Byzantine scholars were even reluctant to accept Islamic astronomy, despite the fact that it was itself founded on ancient Greek astronomy. Science for the Byzantines meant Greek science alone.

Nicaea lacked teaching materials, books, and libraries. In order to study, aspiring scholars traveled from town to town in search of teachers and their books. One of the tasks of the emperors of Nicaea was to remedy these deficiencies. The most obvious solution was to collect the scattered manuscripts and recopy them or, even better, to reedit them. Reediting texts meant rereading them, and rereading them renewed Byzantine science. This led to a second humanistic movement in Byzantium during the Palaiologan dynasty, which was the pure product of Nicaea. The effects of this movement proved much more profound than those of the first humanist movement in the ninth century. This time—after the shock of the fall of the imperial capital in 1204, which marked the end of nine centuries of political continuity—Byzantine society was ready to accept change. Thus, by the end of this period, Byzantines again possessed most of their scientific literature from before the Latin conquest, some of it reedited.

Change came on two fronts: in the religious domain of the Orthodox Church, which demarcated Byzantium from its Latin conquerors, and in the secular domain of ancient Greek knowledge, which demarcated Byzantium from the rest of the *oikoumenē*. An essential trait of the shift in mentalities was the gradual acceptance of the scientific knowledge of other civilizations. This assimilation provoked considerable debate throughout Byzantine society over the validity and utility of non-Hellenic science. As we shall see, the famous Hesychast debate of the fourteenth century would center on whether alien knowledge was worth the trouble of acquiring it. Such debates, however, would largely end by the close of the fourteenth century.

Nicephorus Blemmydes: Scientific Monk

Although Michael Psellos was the most widely known Byzantine scholar in the West (and still remains so in the historiography of Byzantium), in the Orthodox world he was eclipsed by Nicephorus Blemmydes (1197–1272). Blemmydes' personal story illustrates a number of facets in the relationship between science and religion in thirteenth-century Byzantium. When his parents fled Constantinople in 1204 for Prusa, he was seven years old. In Prusa, he learned the *hiera grammata* (sacred letters—i.e., learning to read and write with sacred books as

For that reason, they are included in heaven. For only the divine is unlimited, indivisible, perfect in itself, and infinite."[7]

Blemmydes engaged in a discussion on the nature of spirits (souls). For him, souls are material. What is the nature of their matter? He did not say but was certain that it had nothing to do with the four known elements from which perceptible bodies are composed. For these latter, he followed the Aristotelian tradition that they are always composites, because the simple elements do not exist singly in nature. Moreover, even if they existed, they would not be perceptible by our senses, for only composite bodies can be perceived because only these have a form; simple elements are perceptible only by the mind. The elements that by definition ought to be simple bodies are not really so, for they derive from a primary matter that can assume (according to added qualities) the properties of known elements. This thesis is reminiscent of the primordial matter of Gregory of Nyssa, from which the four elements come (see chapter 1). Although Platonic in his inspiration, Blemmydes is very much a materialist: everything that was created by God is matter.

But Blemmydes did not merely repeat arguments taken from his predecessors; he developed his own theory with a clarity that departs from their conjectures. For him, celestial bodies are composite bodies and not made of ether, because they are perceptible by the senses. In effect, the heaven of the fixed is visible and palpable. The visibility of a body is due to its colors, and the colors of heavenly bodies are due to the clarity of fire. This proves that fire is part of their composition. Then as these bodies interact, they are palpable, and this property belongs to solid bodies such as the earth.[8] Now, because the composition of heavenly bodies includes fire and earth, which are the two extreme elements (the lightest and the heaviest), it follows that the intermediate elements (air and water) also enter into the composition of these bodies. Moreover, these latter two elements are indispensable for linking the two extreme elements. Although the stars are constituted of all four elements, the preponderant element is fire, a very Platonic idea. But it is not only the stars that are composite bodies. The first heaven that is invisible is also such a body, for it, too, is perceived by the senses. Indeed, this heaven is the cause of diurnal rotation, and this movement is perceived by our senses.

What about place and motion? Here too, Blemmydes distanced himself from Aristotle and followed Gregory of Nyssa and such pagan Neoplatonist philosophers of the fifth and sixth centuries as Syrianus, Damascius, and Simplicius. In effect, his theory of movement derives from his theory of place, which is inspired by these three philosophers. Aristotle defined place in relation to the container and not the content: the place of a body at a given instant is the place delimited by the interior surface of the milieu that contains this body. Contrary to this, Syrianus introduced the notion of the relative positions of bodies to define place, and Damascius went further by referencing a three-dimensional space in which bodies move or remain at rest. Blemmydes believed that a body is found in a place, movement in a body, and time in a movement. This interrelation of movement and place led him, in order to determine movement, to determine place in relation to bodies. And although he began by defining place as the boundary of the container ("location is the immobile limit of the container, for it contains the content"), he departed from Aristotle by affirming that, "to be more precise, place is the measure of things positioned, just as time is the measure of things mobile," which reminds us of the system of measurement imagined by Damascius.[9] Making a distinction between *position* according to Damascius and *place* according to Simplicius, Blemmydes thought there are two kinds of position, essential and alienated, and two kinds of place, natural and factual. There are also two sorts of natural place: that which orders the parts of a body, and that which positions each body in relation to the whole. To take the example of the earth, if it moves away from the center of the world, it will keep its integrity (hence the essential relationship of its parts) but it will also tend toward the center of the world (to regain its essential position in relation to the whole).[10] This contradicts Aristotle (but not all Aristotelians), who affirmed in *De caelo* that if someone tried to move the earth from its place, it would remain immobile.[11] Finally, Blemmydes concluded, "bodies move [he is speaking of natural movement] to occupy their own place, and immobile bodies to occupy the place of the location that is their due and that they love. For this reason, place is not the boundary of the container."[12]

Blemmydes went on in this extensive discussion of place and the motion of bodies to lay the groundwork for his thoughts on the nature of heaven. Blemmydes' goal was to demonstrate that the heavens do not need the perfect and imperishable element (ether) in order to move naturally in a uniform circular motion. According to Aristotle, whereas the natural motion of the four elements in the sublunary world is rectilinear, that of the fifth element is a uniform circle, because a perfect element must necessarily move in a perfect motion. Ether is perfect because it is incorruptible, and uniform circular motion is perfect, too, because it is absolutely symmetrical. Blemmydes did not follow Aristotle but he seems to have been inspired by Simplicius, who, in his *De caelo*, transmitted Xenarchus's theses (first century BCE) on the natural movement of bodies: rectilinear motion is the natural movement of the four elements not in their *natural* state but in their state of *becoming*. This means that these elements, when they time is a notion of the created world (and hence "before" does not exist outside creation). The Eastern Church often employed the notion of the nonexistence of time outside the created world, especially in its discussions with the Catholics over the existence of purgatory (see chapter 8).

In discussing God's omnipotence, Blemmydes invoked an argument extremely rare in the theological literature of the Eastern Church: the idea of the plurality of worlds. Addressing those who wondered about the omnipotentiality of God in the event the world is not coeternal with him, he wrote: "They are asked 'if God is omnipotent, would he not be capable of producing not only this world but many others?' If he were not capable of producing other worlds, how could this omnipotent being of all creative science, who has the power to do anything, not be weak?" Blemmydes deduced that if God did not create other worlds when he could have done so, this showed that he is voluntarily *and* potentially a creator; he creates when he wants to, not when he can.[18]

A spiritual child of the Empire of Nicaea in the thirteenth century, Blemmydes reread Hellenic science and, like a new Philoponus, he harmonized it with religious dogma. The philosophical problems that he posed were not original, but his synthesis of them was quite new. On many points, such as the plurality of worlds, it recalls a similar discussion then developing in the Christian West. Still largely unstudied, the scientific interaction between the Eastern and Western worlds would take place across the south of Renaissance Italy and via discussions between the two churches (see chapter 8).

This outline of Blemmydes' career shows that the church remained omnipresent in secular education during the Nicene Empire. Blemmydes lived as a monk; his school, though financed by the state, was located in a monastery; he advised both emperors and patriarchs and engaged in discussions of church unification. The accusation against him for not following Orthodox dogma seems to have done him no harm, because he was later offered the patriarchate. Throughout his career he collected and commented on Greek scientific texts, usually refraining from imposing his own religious views on them. He never questioned the legitimacy of acquiring secular knowledge of nature, and he wrote a book on natural philosophy that eclipsed previous similar textbooks. His accomplishments greatly aided the emperor's ambition to make Nicaea a thirteenth-century Athens.

CHAPTER SIX

Political Debates Become Scientific

The Era of the Palaiologos

The usurper of the throne of Nicaea, Michael Palaiologos (c. 1224–82), proved to be the last great Byzantine emperor. Three years after Michael dethroned the legitimate John IV Laskaris and assumed the title of emperor of Nicaea, in 1258, his army managed unexpectedly to retake Constantinople (on 25 July 1261) and oust the Latin emperor, Baudouin II. As emperor, Michael VIII deployed all his energy and diplomatic skill to restore the city to preeminence and to remove the two dangers that threatened the Byzantine Empire: a crusade by the Latins to retake Constantinople and the Turkish advance in Asia Minor. Michael thought that only one means could ensure the survival of the fragile empire: the union of the Orthodox and Catholic churches. By winning the support of the pope, Michael aspired not only to end the Latin crusade against his empire to the East but also to forge an alliance with the Latins for a joint crusade against the infidel Turks.

Michael inaugurated a policy that would last until the ultimate fall of the Byzantine Empire. Negotiations with the Catholic Church proved interminable and resulted in ephemeral unions, the first concluding with an ecumenical council in Lyon (1274). Michael and most of his successors tried to persuade the ruling class of the validity of such a union, but the lower clergy and the general populace refused to submit to a pope or to accept the contentious Western teaching that the Holy Spirit proceeded from both the Father and the Son (as opposed to the Father alone, as stated in the original Nicene Creed). Scholars participated actively in these ferocious debates and sometimes served as theological experts in negotiations with the Catholics. Often their careers depended on the respective fortunes of the unionist and anti-unionist factions.

In reconquering Constantinople, Michael wanted to restore Byzantium to its lost splendor. (Once Europe's largest and richest city, Constantinople itself had shrunk to thirty-five thousand inhabitants under the Latin rulers.) Along with

The combination of theology and science at Pachymeres' school contributed to the scientific training of a significant number of church dignitaries and also to the valorization of science in the collective awareness of the Byzantine aristocracy. The vogue for astronomy, not only for the utilitarian purposes of astrology and calendar computation (especially Easter) but also for the intellectual exercise of predicting eclipses by complicated calculations, is an example of the value of science, which would lead to a proliferation of scientific texts without precedent since the end of antiquity. However, this increased visibility of secular texts circulating in the church would soon lead to reactions and debates associated with the famous Hesychast polemic.

Another clerical scholar, contemporaneous with Pachymeres, was Maximos Planoudes (1255–1305). Born in Bithynia, he was probably an autodidact. He established himself in Constantinople after the retaking of the city by the Byzantines in 1261. Close to Michael VIII, he was a partisan of church union. With a good knowledge of Latin at a time when this language was despised by Orthodoxy, this admirer and connoisseur of Latin culture translated Augustine's *De Trinitate* in order to contribute to the discussion of the procession of the Holy Spirit. Under the anti-unionist Andronicus II (r. 1282–1328), he gave up his unionist enthusiasm and thereafter remained neutral on the issue. Around 1283 he took the monastic habit under the name Manuel and withdrew to teach and write in an unidentified monastery. Late in the century, he took part in a diplomatic mission to Venice, where he remained for some time, exposing Venetian intellectual circles to secular Byzantine culture.[8]

As a pro-Latin humanist, Maximos the monk remained open not only to Western scientific culture but to all of "barbarian" culture. His most important contribution to science was his book *Indian Calculation*, which he wrote around 1300.[9] There he introduced his Greek readers to Arabic numerals, including zero, as well as to methods of calculation that he called Indian. Arabic numerals had already appeared in Byzantium on the margins of a ninth-century codex of Euclid and in an anonymous treatise published around 1252.[10] But it was only after the appearance of Planoudes' book that Byzantine mathematicians started to use them, albeit timidly. Despite the evident practical utility of Arabic numerals, most Byzantines remained faithful to the Greek tradition of alphabetical numbering and to the hexadecimal system.

Astronomy against Physics, Metochites against Choumnos

The fact that the educational milieu was mixed with that of the aristocracy (notably imperial administrators) sharpened the confrontations that inevitably appeared among the officers of a shrunken state with a dwindling number of important posts. Within this small circle of officers of the Palaiologos empire, in addition to debating the fundamental ideological issue of rapprochement with the Catholics, rivalries for posts were never lacking. Throughout Byzantine history, reasons for conflict among state functionaries had always been abundant: intrigues over succession, fierce dogmatic debates, personal ambitions. The new ingredient was science. The enhanced place of science in the Byzantine collective awareness, the emperor's sponsorship of higher education, the secularization of the Patriarchal School, and the fact that rivals often had received similar educations all helped to explain this new phenomenon. A striking example of a scientific debate that concealed a political rivalry was the famous polemic between Theodore Metochites (1270–1332) and Nikephoros Choumnos (c. 1260–1327), which bore on the mastery by each scholar of the natural philosophy of Aristotle and the astronomy of Ptolemy.

Choumnos had studied at the imperial Philosophical School of Constantinople under George of Cyprus, who gave him excellent instruction in Aristotle's natural philosophy but only a basic introduction to astronomy. Supported by his master, Choumnos took part at an early age in an embassy to the Mongols and later climbed the rungs of the state ladder. As adviser to the emperor and his *mesazōn* (chief minister), he became very rich; at his political apogee, in 1303, he married his daughter Irene to John Palaiologos, the son of Andronicus II. Around 1309, he was downgraded to prefect of Thessalonica and then appointed *epi tou kanikleiou*, the prefect of writing (a sort of secretary of state). He remained influential concerning theological and political questions such as church union. During his political career, Choumnos continued to write about various things, including natural philosophy with an Aristotelian influence.

Theodore Metochites was the first great scholar of the Palaiologos era who had not known the Empire of Nicaea. Born in Constantinople, he learned to read from his father, a fervent unionist, before he was exiled by the successor to Michael VIII, Andronicus II, who had given up on the union of Orthodox and Catholic churches. Metochites continued his studies, probably in Nicaea, where he attracted Andronicus's notice for his eloquence and erudition in the course of his visit to this city in 1290. Henceforth, he would climb the ladder of power, holding various *logothetēs* posts and leading embassies to the Latins, the Serbs, and the Armenians. Around 1305, he became *logothetēs tou genikou* (finance minister) and *mesazōn* to the emperor. Becoming very rich and close to the emperor, he rose to be great *logothetēs* and married *his* daughter Irene to a *different* John Palaiologos, this one a nephew of Andronicus II. Like Choumnos, he remained very cautious about expressing his position on church union.

In 1313, at the already advanced age of forty-three, he started to study astronomy, encouraged by the emperor, who highly esteemed this science. As a youth, Metochites had lacked the opportunity to pursue his studies in the schools of the capital, and he thought he was deficient in mathematical knowledge. But now he brought into his home the astronomer Manuel Bryennios (c. 1275–c. 1340), reputed to be an expert on Ptolemy; with his help, Metochites studied Ptolemy's *Grand Mathematical Syntax* (the *Almagest*) and also Euclid and Apollonius. Metochites quickly became an expert in Ptolemy and around 1316 wrote the most voluminous Byzantine astronomical textbook, the *Elements of Astronomical Science*, in which he explicated the *Almagest* in a rather complicated manner.[11] The fact that Metochites was capable of composing such a book after only three years of study (and in parallel with his important functions as *logothetēs tou genikou* and *mesazōn*) shows either that he was already well advanced in astronomy before beginning his work with Bryennios or that the latter helped him in writing the book. Either way, Metochites achieved great fame as an astronomer and presented himself as the restorer of Ptolemy's astronomy and the heir of Greek astronomy, which he considered the queen of the sciences. He stated with pride: "I predicted the precise circumstances of solar and lunar eclipses. Thus I affirmed in the eyes of the general public that this is a true science, powerful and exact."[12] Metochites hoped to purify it of Oriental—that is, Islamic—influences and clearly demarcate it from astrology.

Choumnos and Metochites had every reason to get along—but also to hate each other. They were of the same generation (Choumnos was at most ten years older), rich, well educated, interested in the sciences, close to the emperor, important dignitaries—and neutral in the great ideological conflicts of the day. In fact, they esteemed each other. Their correspondence about their respective books was friendly in tone and even full of praise. Metochites declared himself "ready to read the books of this noble and fecund author Choumnos," while the latter asserted that Metochites' opinion was worth a thousand others. On one occasion, Choumnos humbly asked for Metochites' opinion of his book on the atmosphere and compared the eloquence of the Byzantine savant to that of Homer. In return, Metochites remained mute with admiration before the excellent works of Choumnos. Their relationship continued this way—until the moment when

their respective ambitions brought them into competition for primacy in imperial favor.[13]

The controversy began around 1305, when Metochites, returning from Thessalonica where he was prefect, replaced Choumnos as *mesazōn* to the emperor.
This was a little before he managed to marry his daughter to Andronicus's nephew
to seal the emperor's friendship and trust—and counter the fact that Choumnos's
daughter was already an imperial in-law. The ambitious Metochites had finally
succeeded in obtaining a very high rank in the hierarchy. The rank of people at
court depended on their official function. Before Metochites, the great *logothetēs*
occupied the eleventh rank, whereas the prefect of writing, Choumnos's office,
was only thirteenth. As soon as Metochite became great *logothetēs*, the rank of
this office rose to ninth position, four ranks more important than Choumnos's.
The latter tried to regain Andronicus's favor and recover his rank among the
imperial dignitaries. He found allies among other dignitaries jealous over the
rapid ascension of Metochites, who had formed an anti-Metochites party, led by
Choumnos and Theodore Palaiologos, son of the emperor. Metochites had vis-
ibly enriched himself from his relations with Andronicus and enjoyed the fame
of being a great savant.

Choumnos's weapon against the great *logothetēs* was not political argument
or even calumny, but scientific critique. Now he transformed Metochites from
a great scholar of unequaled eloquence and limpidity into a scientific ignora-
mus, dreary and convoluted in style. The latter parried the attack by dismissing
Choumnos's polemic against Aristotle as foolish, noting also that he had made
the gross astronomical error of inverting the order of Saturn and Jupiter. And
thus began a great debate over the philosophy of nature and astronomy.[14]

When Choumnos argued that the study of the philosophy of nature constituted a means of understanding the world of the Creation (to the extent it was
linked with theology), Metochites responded that Creation could not be ap-
proached by the sciences. Theology ranked above mathematics, and it was accessible only by instruction from God. As regards science, it was evident that
mathematics should be preferred to physics and other subjects.[15] Unlike the
church fathers (see the *Hexaemeron* of Basil), who distrusted the Greek philosophers and pointed to their contradictions, Metochites, like other Byzantine humanists of the Palaiologos era, posed as the spiritual heir of these philosophers.
But when he tried to present Greek philosophy as a single system and to force
an agreement between Plato's cosmology and Ptolemy's, Choumnos thought he
had found his opponent's weak point—and evidence that the great *logothetēs*'s
knowledge was not in fact superior to his own. Then Metochites tried to "cover"

Plato's ignorance by falsifying the writings of the great philosopher. The problem was the following: Platonic cosmology posited that the seven planetary spheres move from west to east, and the eighth sphere of the fixed stars is responsible for the diurnal movement and moves from east to west. Since the discovery of the precession of the equinoxes (Hipparchus, second century BCE), the problem was posed of the sphere responsible for this west-east movement. The simplest solution, followed by Ptolemy, consisted of adding a ninth sphere that goes from west to east. Yet the order of the spheres responsible for diurnal movement and for precession was often inverted, as Metochites had done when he postulated that the eighth sphere turns from west to east. Choumnos noticed the contradiction with Plato, who made this sphere turn in the opposite direction. Metochites' answer could have been very simple and convincing if he admitted that Plato, not knowing about precession, could not imagine the eighth sphere moving from west to east. Instead, he falsified the writings of the great philosopher, by adding to Plato's text just one letter, a sigma, producing σεπτάς instead of επτά—which means "venerable" instead of "seven." Thus, Plato supposedly did not give a precise figure for the "venerable" spheres turning from west to east, which could perfectly well be eight in number.

Before long, the two high Byzantine dignitaries, Choumnos and Metochites, were indulging in a perennial scientific debate: Was it sufficient to "save the phenomena" or did one need also to explain them? Since antiquity, the Greek fathers had given the answer that suited good Christians: first causes would remain inaccessible, for no savant could interpret the creative will of God; such questions belonged to the domain of theology. Nevertheless, it was good to try to understand the functioning of the world, for this made humans admire Creation even more. In the tradition of the Hexaemerons, from that of the Jewish Philo of Alexandria through that of Philoponus, God appeared as a surveyor.

Choumnos chided Metochites for claiming to be an astronomer when the zephyrs of the Pontus (the south coast of the Black Sea) prevented any observation of the sky from there. He himself posed as a physicist who demanded observational verification of his theories. However, most Byzantine scholars found this position unconvincing because few of them were interested in the "practical" side of science, namely, experiments and observation. This bias explains why, unlike contemporary Arabs, they left almost no scientific instruments or accounts of observations of the celestial phenomena that they themselves had predicted. In effect, the only Byzantine instruments that have been conserved to our day are an astrolabe of Persian inspiration, constructed in 1062, and fragments of another astrolabe. Although smitten with astronomy, the Byzantines were prejudiced against observation for two reasons. First, they considered themselves the sole and legitimate heirs of Greek science; thus, Ptolemy was *their* astronomer. Second, the influence of the rational sprit of antiquity, especially Plato's, reinforced by knowledge of the world derived from sacred texts, gave them the feeling that observation was a servile and illegitimate thing and that using imperfect instruments was inferior to pure reasoning. Byzantines may have written many treatises on the astrolabe, an instrument associated with astrology, but this provides little evidence of a culture of observation; instead of emphasizing the importance of observation, these treatises featured the drawing of astrolabes merely as the occasion for a display of fine geometry.[16] In his last triumphant speech against Choumnos, Metochites enumerated the tasks of the true savant: to know how to calculate solar and lunar eclipses, the planetary movements and especially their retrogradations, the rising and setting times of stars, the occultation of stars by the moon—everything founded on the *Almagest*, which provides the "redoubtable and absolutely inexpugnable ramparts of geometry."[17]

In this context, it is not surprising that Metochites, posing as an expert in theoretical astronomy, the queen of sciences, ultimately won the contest against Choumnos. Eclipsed by Metochites, whom he considered an *arriviste*, Choumnos lost the privileges that had enabled him to get rich. Embittered and ill, he no longer appeared much at the palace. He died in 1327. His battle with Metochites, ostensibly fought over astronomy, ultimately had major political consequences, playing a role in the ensuing civil war between Andronicus II and his grandson Andronicus III. The great *logothetēs* himself, Metochites, soon saw his fortunes decline—literally, since his goods were confiscated. He became a monk and, taking the name Theoliptos, retired to the monastery of Chora, which in the time of his good fortune he had renovated and endowed with a fine public library and a school.

Greek or Latin Reading of Aristotle: Gregoras against Barlaam

A few years after the Metochites-Choumnos controversy, a similar scenario played out between Nikephoros Gregoras and Barlaam of Calabria. Nikephoros (c. 1295–1360) was born in Herakleia in the Pontus. An orphan, he was, thanks to the help of an uncle who served as the metropolitan of Nikephoros's native city, able to receive a good education from John Glykys (future patriarch John VIII) and Theodore Metochites. Metochites subsequently named him tutor to his children and professor at the school of the monastery of Chora. Entering into the circle of Andronicus II (thanks to his teacher), he became around 1322 *char-*

tophylax (archivist). The emperor, a lover of the sciences, eagerly organized lectures in which scholars talked about science in front of court dignitaries. Gregoras became a favored lecturer, valued for his wide knowledge, and he emerged as the preferred professor at the court. A brilliant astronomer, he proposed around 1326 a reform of the calendar. Because of the roughly approximate length of the year (365.25 days) of the Julian calendar used by the Byzantine Empire, the equinoxes were already eight days behind the true equinoxes, something that posed various problems, including determining the day of Easter. Although the context (a renaissance of the sciences, an enlightened emperor) appeared favorable for a change to a more correct calendar, the moment was not propitious. The Orthodox Church, suffering from the shock of the aborted union with Rome and from restlessness among the monks and lower clergy, refused to endorse the proposed reform. Gregoras himself, in his *Roman History*, indicated that his calendar proposal was rejected in order not to "confuse the ignorant and divide the church."[18]

A close adviser to Andronicus II, Gregoras fell into disgrace when Andronicus III won the dynastic battle. It was around this time that the southern Italian Barlaam the Calabrian (c. 1290–1348) arrived in Constantinople. Barlaam, a monk erudite in natural philosophy, took over the post in which Gregoras had formerly excelled, that of court *didaskalos* and privileged adviser to the emperor on religious questions. Gregoras, who hoped to regain imperial favor, his official titles, and his confiscated fortune, could not accept seeing his place usurped by a new man, who moreover was from the country of the "Latins."

The scientific controversy between Gregoras and Barlaam was not merely personal; it also opposed Byzantium to the West and put two conceptions of Aristotle in confrontation: the Latin one and the Byzantine one. Gregoras accused the "Latins and Italians" (targeting Barlaam) of being occupied only with "the wisdom of Aristotle concerning physics and logic" and not understanding his deeper thought, because they were studying the great thinker in the translations of their "poor language."[19] Gregoras, following the tradition of Metochites, considered himself heir of both Aristotle and Plato and advanced the ideas of the latter against the Latin scholastic. Barlaam counterattacked. The pretext was the writing by Gregoras of three chapters on the relations of harmonic tones with celestial bodies, in order to "complete" the work of Ptolemy on the *Harmonics*. Barlaam accused Gregoras of mistakes and also of plagiarism, for allegedly copying one of his new chapters from ancient manuscripts. Gregoras responded by challenging the Calabrian in the domain in which he excelled, astronomy, and particularly the calculation of eclipses.

During the first half of the fourteenth century the residents of Constantinople witnessed a great number of solar eclipses. The growing difficulties of an empire faced with invasions and fractured by interminable struggles for the throne led to a renewal of interest in astrology, a science we know had never been neglected by the Byzantines. Even the enemies of the occult sciences, such as Metochites and Gregoras, had paid attention to eclipses as signs. In fact, the latter had established a link between eclipses and disasters such as the death of his former protector, Andronicus II, or the invasion of the Scythians into Thrace.[20] Such difficult and fastidious calculations not only demonstrated the great skill of the calculator but also performed a service to Byzantine society.

Gregoras's calculation of the eclipse of 16 July 1330 launched a veritable fashion in the world of Byzantine scholarship. Calculating solar and lunar eclipses became a "must" for someone who wanted to appear a savant. The calculations were first based on Ptolemy (Metochites's *Elements* or the *Almagest* itself); later the introduction of Persian tables facilitated the work. For the eclipse of 1330, Gregoras gave the calculation according to Ptolemy's *Almagest* as well as to his *Handy Tables*.[21] Exploiting his success and eager to annihilate his adversary, Gregoras challenged Barlaam to calculate the next eclipse, anticipated in 1331. Later, Gregoras presented this challenge as a public confrontation between him and Barlaam in the home of John Kantakouzenos (a powerful politician who had supported Barlaam in his astonishing ascension at court). Probably the competition was a literary *topos,* a dramatization to demonstrate the superiority of Gregoras, an impression strengthened by the confrontation's taking place in the home of this protector of Barlaam, who in fact failed. But the Calabrian was not definitively beaten; like Gregoras, he successfully calculated the subsequent eclipses of 1333 and 1337.

Gregoras wrote three pamphlets against Barlaam: the *Contradiction* and the dialogues *Concerning Wisdom* and *Phlorentius.* His attacks bear on his adversary's ignorance of astronomical science and on the general ignorance of Latins, who never bother "with either astronomy or any wisdom that has flowered among the Hellenes," whereas in Byzantium letters and rational knowledge "flourish anew and are nourished and improved." Not surprisingly, neither man emerged victorious from this bitter polemic.[22]

The scientific debates between dignitaries of the regime at the start of the fourteenth century illustrate the humanist atmosphere that reigned at the court of Andronicus II and continued under Andronicus III. Although we recognize in the ideas of the protagonists (especially Metochites and Gregoras) the influence of the Orthodox Church, religion appears relatively absent from the debate.

Indeed, it is often difficult to distinguish ecclesiastical influences from such Neo-platonic ideas as the superiority of the mathematical sciences over the philosophy of nature or the distrust of Latin Aristotelianism and of Arabic and Persian science. The actors debated each other in a humanist framework that gave great importance to scientific knowledge, with the victors emerging with sufficient prestige to assure them of successful careers in politics.

CHAPTER SEVEN

True Knowledge and Ephemeral Knowledge

The Hesychast Debate

"The struggle that so often in Byzantium opposed the party of the monks to certain high ecclesiastics sponsored by the emperors was largely based on the aversion among wide sectors of Monachism to the appearance of secular humanism," wrote the great Byzantine historian John Meyendorff, in describing the so-called Hesychast debate that began in the fourteenth century. "This was a veritable drama within Byzantine civilization."[1] As we have seen, this conflict had already run for several centuries and pitted the ruling class, represented by the court and the upper clergy, against the monks, the lower clergy, and most of the general public.

Hesychasm (from the Greek *hēsychia* [quietude]) emerged within the Orthodox Church as a method to reach "true knowledge" of God. A mystical and ascetic practice, it became a way of life and a school of prayer. Hesychasts meditated and prayed, invoking the name of Jesus in order to attain communion with God. Social living and secular knowledge did not interest them; they sought *sōtēria* (salvation) for themselves and for others. In the words of a modern Hesychast, Father Seraphim of the Russian monastery of Mount Athos:

> To pass through the summit of sacrifice is to discover that nothing belongs to "me." Everything belongs to God. It is the death of the ego and the discovery of "true self." Meditating like Abraham means adhering by faith to the One who transcends the Universe, means practicing hospitality and interceding for the salvation of all human beings. It means to forget oneself and to break even the most legitimate attachments in order to discover oneself, those close to us, and the whole Universe, inhabited by the infinite presence of "The One Who Alone Is."[2]

The roots of the Hesychast movement went back to the fourth century, to the Egyptian hermit fathers who went into the desert to confront the devil and to get closer to God. The Byzantine monks had maintained this mystic and especially ascetic tradition, which did not always please the higher clergy and the emperors. Symeon the New Theologian (949–1024) expressed this mystical tendency by seeking through prayer a vision of uncreated light (*aktiston fōs*), the light that is not the one that illuminates the world.[3] For the monks, this vision was achieved only after great spiritual effort; it brought about direct communion with God. This mystical theology of uncreated light (which amounts to believing that one can know God through experience) would be at the center of the Hesychast movement, to be expressed in a dynamic way in the fourteenth century by Saint Gregory Palamas. So a controversy that had blindly opposed monks and the partisans of humanism for several centuries burst into the open, and the pretext was the arrival in Constantinople of Barlaam of Calabria.

Humanism against Hesychasm, Barlaam against Palamas

As we have seen in chapter 6, Barlaam (c. 1300–1350) was a Greek Orthodox believer from southern Italy. Barlaam's involvement in the discussions over the Eastern and Western churches became the pretext for the controversy that opposed the Calabrian scholar and the hermit monk Gregory Palamas, leader of the Hesychast movement. Gregory (1296–1359), son of a senator, orphaned at an early age, had been brought up at the expense of Emperor Andronicus II and studied at the Imperial University under the supervision of Theodore Metochites, great *logothetēs* and scholar. But before finishing his studies, he renounced Hellenic science and withdrew to Mount Athos to become a monk. In his own words, "[I] had forgotten almost all literary science, although due to a pressing need and despite myself, I tried to assimilate something of it, as much as possible." Once at Mount Athos, Palamas argued that the only worthwhile knowledge was theosophy, which could be attained only by purging oneself of secular wisdom: "We monks rise to a unique theosophy, superior to any philosophy," and for this to happen "the very nature of intelligence must be quite naked, a temple of prayer, foreign to any subtlety and any human method."[4]

The idea that man, through prayer and ascetics, could have a vision of God was very popular in Orthodox religion, and miracles continued to play an important role, and have done so right to the present day. Palamas expressed this idea forcefully in an era when the Byzantine people felt themselves under siege from both the Muslim Turks and the Catholic Latins, and so they turned once more to Orthodox religion and especially toward its more mystical side. We saw that the negotiations over the union of churches did not please the lower clergy and the monks, who considered the Catholics as their mortal enemies. But given the influence of these social strata on ordinary people, the reaction against the intellectual renewal brought about by humanism was expressed across most of Byzantine society. This majority felt itself much closer to Palamas than to Barlaam and court scholars, and its reaction would increase the chasm that separated these two expressions of Byzantine culture. Yet Byzantine humanists (who since the ninth century had considered Byzantium as the scientific heir of ancient Greece) also considered the knowledge of other civilizations as inferior, and consequently they despised the West. The Latin conquest of 1204 only strengthened this feeling. A little more than a century later, the Hesychast reaction against Hellenic learning meant that these scholars were starting to envy the Italian side, where scholarly expertise was welcomed with respect. The gap between Byzantine savants and Western savants would gradually narrow, and there was increased receptivity toward Latin culture.

In the era of Palamas, the humanists were represented by Barlaam, who would become the bête noire of the Hesychasts. In order to counter the pope's legates, Barlaam wrote a great number of polemical works that were widely disseminated, as attested by the number of manuscripts preserved. These works treated the question of the *filioque* dogma of the Catholics (that the Holy Spirit emanates from both the Father and the Son, as opposed to the Father alone, according to the Orthodox), as well as the doctrines of Thomas Aquinas, whose scholastic and demonstrative approach was far removed from the Orthodox faith. In fact, Thomism had one thing in common with Palamism: both doctrines argued that the existence of God could be demonstrated, the former by reason and the latter by experience. Barlaam, following Aristotle and Pseudo-Dionysius the Areopagite (late fifth century CE), maintained that theological truths could not be demonstrated.[5] This was the heart of a debate that would extend to the validity of secular science and to the perception of the world by humankind. Barlaam, in the face of the legates' arguments on the *filioque*, opposed his own interpretation of the texts of Pseudo-Dionysius, according to which there was no valid demonstration of truths concerning God. This thesis was as opposed to the reasoning of Saint Thomas Aquinas as it was to the demonstrative *apodictica* of Palamas. According to Barlaam, the only theologians who could approach knowledge via experience were the first church fathers, and since then, all others have been theologians "of opinion" who believe in the revelations of the former. Barlaam (as a humanist) mentions the Greek philosophers who had already made such a distinction between these two categories of theologians and who had defined divine principles as outside human knowledge.

On the theological level, the debate between the humanists and the Hesychasts occurred on three fronts. The first was the meaning of purification, which was one condition of knowledge. Saint Basil in his *Hexaemeron* explained that the soul must be purified by prayer in order to be able to gain access to the truths of the book of Genesis. Barlaam and the Byzantine humanists, in the footsteps of Plotinus, believed that it was the acquisition of vast secular knowledge that leads to the purification of the intelligence. By contrast, the Hesychasts fervently practiced scriptural commandments to obtain the soul's purification. The second front concerned the prayer method. Barlaam had accused the Hesychasts of letting the mere body participate in prayer. Palamas's answer was that the body was the "temple of the Holy Spirit" and must participate in prayer just as it does in Communion. The third front was the unknowability of God, as discussed above.

The front that would involve the whole society and would mark the future of Byzantium and the other Orthodox countries, especially Russia, was the first point, for it led to a debate over the importance of secular science and ultimately over material civilization. The insistence of the humanist Barlaam on citing Greek philosophers (especially Aristotle and Plato) in order to support theological views aroused the wrath of the Hesychast party against the secular sciences, which were already seen in a bad light by the monks and the lower clergy. According to a fervent representative of the Hesychast party, the biographer of Palamas, Philotheos Kokkinos, who was the patriarch of Constantinople twice, from 1353 to 1354 and then again from 1364 to 1376, Barlaam "rejected the rational demonstrations against heresy made by the wise theologians, by calling 'divine' and 'illumined by God' the sages of the Greeks: Aristotle, Plato, and all those in their company. Against such prattling, prejudicial to piety, the noble Gregory Palamas rose up with words worthy of him and the truth."[6]

In fact, Barlaam (in the wake of Philoponus) did exalt the sages of antiquity, for according to him they too had benefited from divine revelations (Philoponus even thought that Aristotle and Plato had read the Hebrew Bible). Addressing Palamas, he wrote:

> You appear to denigrate the Ancients, who forbade demonstration in the domain of divine things. But I do not see why I should not consider them admirable, they who so nobly recognized human powerlessness and the transcendence of God; and when I see them include among the elements interior to the soul the methods of demonstration and analyses . . . and declare that they prefer in the domain of material and natural things the reasoning that suits

these things, and when in the domain that surpasses us I see them assert that the blessing of a vision from above and the burst of an intelligible light that enables them to unite with divine things, to contemplate the Transcendent better than by demonstration . . . , then I cannot prevent myself from presuming that they too were to a certain extent illuminated by God and so surpassed the majority of men.[7]

Barlaam, going even further than Philoponus, affirmed that, just as the Apostles and the church fathers had, the Greek philosophers benefited from divine revelations that made them the "theologians of experience" who enabled those of us who follow them to have "opinions." The classic writings are thus not simple philosophical opinions; they are references for Christians and therefore have the right to be cited on the same level as the writings of the fathers in theological debates.

Such opinions appeared extreme and sacrilegious to Orthodox zealots. Nevertheless, the relation of the Orthodox Church to secular science (called Hellenic in this era) is more complicated than a division between a caste of monks, which rejected it, and a humanist higher clergy, which accepted it.

The Hesychasts, at least the most eminent among them, did not actually reject secular learning, for they continued to consider it useful for understanding and interpreting Creation. They simply believed that this wisdom was not important, because true wisdom (that which brings humans close to God) is found in Hesychastic practice. Palamas himself was a follower of Saint Basil when it came to Creation; consequently, his conception of the world was based on this oft-denigrated Greek philosopher. Philotheos Kokkinos, although attacking the "sages of the Greeks," displayed in other texts an admiration for Aristotle as a scholar. Profane learning was completely rejected only by the humblest monks, who had no contact with higher education, as Palamas or Philotheos did. In fact, the absolute rejection of science was determined by social class. The Byzantine dominant class accepted it, either with fervor (in the case of the humanists) or under certain conditions (in the case of the Hesychasts), whereas the poorer social strata rejected it as useless, never having had much contact with it.

Rather than an attack on secular learning, Hesychasm was a response by the monks to the tendency to secularization within the Orthodox Church, which was the consequence of two factors: the development of Byzantine humanism, which had as a consequence the training of the upper clergy in secular science, and "Caesaropapism," which placed the church under the authority of the emperor. In the Christian West, an analogous phenomenon, the upgrading of secular knowledge, especially fostered by the nominalist attitude that cast doubt on ecclesiastical "mystery" (i.e., the real presence of Christ in the church), together with the secular way of life in the Vatican, would constitute the basis of the future Reformation. In the East any hint of reform was blocked by the Hesychast reaction, which was fighting on two fronts simultaneously: against secularization and for control of the church through its internal hierarchy (not through political power). As a result of the Hesychast movement, the power of the monks in the church increased proportionally to the decline of the emperor's. Henceforth, any effort at union between the two great churches, even if promising in the short term, was doomed in the medium term to failure.

Gregory Palamas's Vision of the World

Gregory Palamas, originally trained in the spirit of Byzantine humanism, including Hellenic logic and science, later combated this same humanism with his own tools. He did not object to the deductive syllogism known as the *apodictic*—on the contrary, he applied it to theology. But whereas with respect to nature he observed that the generalization of our knowledge through experience could lead us to erroneous results, he thought that the apodictic syllogism was infallible with respect to dogma. Dogma cannot admit dialectical thought; it must be clear and stable. How can we reach this certitude? By applying logic and deduction based on the sacred texts that embrace Holy Scriptures and the writings of the church fathers. God presented himself to the world and was materialized, and therefore man can indeed approach God, simultaneously by the mystery employed for spiritual things and by logic employed for material things. It goes without saying that a person who does not have the grace of God (i.e., a humanist) cannot apply apodictic syllogism successfully.[8] Palamas was aware that his use of reason and deductive logic required a defense. "Are learning and the science of discourse bad things?" he wondered. "Of course not, since God has given us science and methodology. Therefore it is not they that are wrong, but their wrongful usage by sinners."[9]

Similarly, the created world can be understood and explained only by those who have grace—the Hesychasts. Aristotle, and the other Greek savants, though realizing that nothing is created from nothingness and that nothing will disappear completely, came to the erroneous conclusion that the world was not born and will never die. Therefore, they deduced something incorrect though starting from a correct realization. To arrive at a true image of the world, experience is not sufficient; one needs the illumination that is granted only to those who believe in the mystery of the church and, through it, enter into communion with God.[10]

According to Palamas (and contrary to the letter of scripture), Father, Son, and Holy Spirit created the world together. This world was actually created in six days, and the seventh that followed was longer than the others because it comprised the whole era that began with the last day of Creation and terminated in the crucifixion and death of Christ. The Resurrection marks the start of the eighth day, which we are traversing now and which will endure until the Last Judgment. This judgment will take place on a Sunday, which is the privileged day because the first day of the week is comparable with the first day of Creation. Palamas contributed also to the discussion by Philo, Basil, and others of why Moses should have called the beginning of Creation "day one" and not "first day"—quite simply in order to make a distinction between them.[11]

An admirer of Basil, Palamas followed the cosmology of the school of Alexandria. Regarding the angels, his ideas were close to those of Philoponus, despite the fact that their conceptions of science were diametrically opposed. Philoponus, as I have already mentioned, was followed enthusiastically by the Byzantine humanists; he considered that the learning of the Hellenic philosophers was valid because they were illuminated by knowledge of the Bible—although similar ideas were truly sacrilegious in the eyes of the Hesychasts. According to Palamas, angels were created before the world, and so they are incorporeal and do not take part in the functioning of nature (as followers of the school of Antioch maintained) but serve for the salvation of humans.[12] Palamas cited Saint Basil's comment that angels are found amid uncreated light; they can traverse the firmament as light does.

The revelation of uncreated light to the Hesychasts was an opportunity to debate the nature of starlight and especially Saint Basil's ideas on this subject. We recall that Basil considered that the light that would illuminate the world existed before Creation, and therefore it is uncreated light. The world was isolated from the light by the firmament, and at the command *fiat lux* it traversed the firmament and lit up the world (see chapter 1). This explanation, which was completely revised by Gregory of Nyssa, who gave corporeal characteristics to the light of the world, is truly problematic, because it introduces into nature an uncreated element, and also because it posits that a created element, the firmament, can arrest uncreated light. This is how the leader of the anti-Hesychasts, Akindynos, posed the question: How is it possible that uncreated light is prevented from traversing the firmament, while the angels do traverse it?[13] Although Akindynos was an adversary, Palamas could only concede to the argument that uncreated light is everywhere and no material wall can stop it. However, it cannot be perceived by the senses, except by a few of the happy elect who have made the superhuman effort of prayer and devotion.[14] It follows that the light that shines on us is not the uncreated light but rather the light discussed by Gregory of Nyssa.

It would be a mistake to see the Hesychast movement (especially its leader Palamas) as hostile to secular learning as such. Palamas was interested in secular knowledge, notably that which described and explained Creation; he proceeded by deductive reasoning based on sense perception. But we have seen that this method was not sufficient for him because it was likely to lead to erroneous conclusions. In order for knowledge based on experience to be valid, it must follow the interpretation of Creation given by the church fathers, especially Basil. But—and this is particular to the Hesychast movement—the world in which we are living is not composed of physical reality alone. According to Palamas, to limit man to perceiving merely the created world would be to condemn him to spiritual misery. A Christian is open to another world that was not created by the imagination of Hellenic philosophers—namely, the uncreated world of spiritual powers. Man may take part in both worlds, created and uncreated, for he is composed of both corporeal matter and an incorporeal soul. God, creator of corporeal and incorporeal worlds, is inaccessible to man in essence but accessible through his actions. This participation in two worlds is the very essence of the Hesychast movement and explains the fact that, despite its followers finding themselves at loggerheads with the humanists, they tolerated secular learning and sometimes even considered someone who possessed it as privileged. The fervent Hesychast Philotheos Kokkinos cited the great humanist scholar Metochites, who was supposed to have said of his pupil Palamas on the occasion of a discussion of Aristotle's logic in the presence of the emperor: "And I believe that if Aristotle were present, he would have made an elegy as good as mine. I maintain that this is how the nature and soul of those who avoid chatter should be, just as Aristotle thought and wrote at length."[15]

What matters most to Palamas is precisely to show that the ancient philosophers, despite the fact that they described the physical reality of the world, were not able to do so completely and exactly, for they could not accede to the true wisdom that is offered only through the methods of Hesychasm. More than being simply ignorant compared to Christians, Plato, Socrates, Plotinus, Proclus, and Porphyrus were under the influence of the devil. Socrates, although judged to excel in wisdom, was possessed his whole life by a demon who had convinced him. For this reason, he taught things contrary to true wisdom, as with his cosmology or, still worse, his ideas on the soul of the world, at least as presented by his pupil Plato in *Timaeus*.[16] As for Plotinus, according to legend a dragon appeared from under his body at the moment of his death, and so Palamas concluded that hidden behind Plotinus's wise teaching was the Father of Falsehood, the devil.[17] The myth that Proclus had a vision of Light gives Palamas the opportunity to argue that it was the work of the demon—the same one that left his head after his death.[18] It is notable that nowhere does Palamas imply that Aristotle was possessed by the demon.

This false wisdom of the ancients is overcome by the spiritual wisdom of Orthodox believers. It is by no means necessary for someone to rise to saintliness for him to be compared to the Hellenic sages: "Not only is the fact of truly knowing God (to the extent permitted us) incomparably superior to the wisdom of the Hellenes, but also knowledge of the place occupied by mankind near to God surpasses all their wisdom."[19] According to Palamas, God has shown us that profane learning is false. But how can any learning conceived by the human mind, a creature of God, be a sin? Ah well, quite simply because this mind is moving away from its real purpose, which is knowledge of God.[20]

As a result of his education by Metochites, Palamas was adept at Greek cosmology, thanks to which he adopted arguments from Basil's *Hexaemeron*. But in certain cases he departed from Basil, developing his own (often contradictory) ideas. Coming to the question (that had been debated since antiquity) of the place of the world and its possible movement, he explained that there is no reason to believe that a space outside heaven cannot exist. On this point, he came into contradiction with Basil, who thought that space was created simultaneously with time and matter, and therefore it involves Creation alone, outside of which nothing exists. Palamas explained that God fills everything and extends to infinity, and within this infinity the world was created. Because nothing prevented the creation of space within the created world, then nothing prevents the creation of space outside of it. So then, why could this world not move, why is it constrained to turn in place around itself? There, Palamas gave two contradictory explanations in the same paragraph. He explained first that "the body of heaven does not extend higher because this higher [the breadth of heaven] is lighter than it; this is why it [heaven's breadth] is above the sphere of ether, by its nature," and then just afterward he asserted that "heaven does not advance upward, not because there is no space above it, but because no body is lighter than it." Finally, he ended by asserting that there is nothing above heaven, not because no space exists there, but because heaven includes all bodies and there can be no body outside it.[21]

But since there is no obstacle, why does heaven not ascend but instead moves cyclically? Well, this heavenly body is much lighter than all the others, hence it is located at the surface of other bodies. At the same time, it is more mobile than the other bodies, and since it has a tendency to move but cannot by its nature separate itself from the bodies above which it is located, it moves constantly around them; and this is not because it has a soul, but because of its material nature. Palamas gives the example of winds that move without rising upward, not because there is no space above them but because what is above is lighter. In all these explanations, we perceive the vague influence of Hellenic culture that incorporates Aristotelian ideas of the natural place of heavy and light bodies but, at the same time, cannot conceive of any notion of symmetry and insists on seeing infinite space as having an "above" and a "below."

If Palamas had been forced to choose among the Hellenic philosophers the one who was closest to the truth, he would no doubt have chosen Aristotle. Our opponent of Greek philosophers cited his ideas countless times as reflecting the reality of Creation. Against the Platonic idea of the soul of the universe, he cited Aristotle in arguing that the soul is the vital force of an organic body that has power in living. For a body to include organs, it has to be composite, and heaven is a simple element.[22] The world according to Palamas (explicitly citing Aristotle) is made up of five elements in equal quantities. But the space occupied by these elements is in inverse proportion to their density. This is why water is more extensive than the earth, the air is more extensive than water, and so on for fire and ether. He asserted that the Hellenes neglected this fact, and consequently they overlooked that nine-tenths of the earth is covered by water. But if the spheres of the elements were concentric, then the whole earth would be covered by water. Therefore, the aqueous sphere is excentric, and Palamas proposed to find its center: manifestly it is not above out heads, for we see that the surface of the water is below us. Consequently, it is below the center of the earth. So it is a matter of determining the size of the spheres of the earth and of water (referring to the element earth, which here is confused with the planet Earth). Knowing that the surface of the sphere of the earth is one-tenth the size of water's, Palamas calculated the size of the radius of each sphere. By these geometric demonstrations, he said, a sphere that has double the diameter of the other has a surface eight times greater, which is valid, in effect, since the surface is proportional to the cube of the radius. From this, Palamas deduced that the sphere of water has a diameter double that of the earth. As in all his demonstrations, the scholar-theologian remained approximate; he was content with this solution—although he had previously asserted that the surface of the earth is more or less a tenth that of water.

By developing this theory of earth-water proportionality, Palamas constructed a very interesting world system, which he even illustrated with a drawing.[23] Since the sphere of water is almost adjacent to the earth's, the latter is inscribed in the aqueous sphere whose center corresponds to the point opposite the adjacent point. As in his argument for the world's movement of translation, here, too, there is an above and a below, with the lower point of the earthly sphere corresponding to the center of the world, while, on the upper part, the sphere of water is conjoined to a tenth of the sphere of earth, because the inhabitable part of the earth corresponds to a tenth of its circumference. Moreover, because the great part of the earth is included in the sphere of water, it becomes evident why there are so many subterranean waters. Because only the upper part of the earthly sphere is free of water, it follows that the antipodes cannot be inhabited. According to Palamas, on this point the Hellenes were also mistaken: there is only one *oikoumenē*, and it is ours; consequently, there is only a single race of humankind.

Although Palamas firmly condemned Plato, he oscillated between this philosopher and Aristotle, and he was even on occasion labeled by Barlaam as Platonizing. In general, we may detect the influence of Plato on his theory of knowledge and that of Aristotle on his physics. Approaching Plato, Palamas explained that man perceives the world though the senses. But he said that what is perceived is not the objects themselves but their copies, which exist independently of reality, for we can represent these imaginary objects at any moment.[24] Approaching Aristotle, he posited a world of five elements, of which the fundamental bodies (heaven, fire, air, earth, and water) are pure.

Palamas came back several times to the power of observation and logic to understand the world: "It is by the intellect that we collect with our senses and our imagination not only what relates to the Moon, but also to the Sun and its eclipses, and the parallaxes of other planets in heaven and their measurements, as well as the constellations, and in general everything that we know of heaven and all the causes of nature, all the methods and the arts."[25] But where does our knowledge of God come from? And of the world itself? It is by the teaching of the Spirit, from which we have learned things about Creation that are inaccessible to the intellect via experience. By the teaching of Moses, hence by the Spirit, we have learned that in the beginning there were heaven and earth. This earth was mixed with water, and these two elements produced air. Heaven was filled with lights and with fires. Contrary to those who claim that matter preexisted Creation, God created the receptacle that carried the potential for all the beings of this Creation.

This insistence on a point that had been resolved long before, the non-pre-existence of matter, shows how the Hesychasts were manifestly worried that the humanists might (out of their love for the Hellenes) defend materialist positions.

This was not in fact the intention of humanists, for in the history of Byzantine science such a position had never been held. The leitmotif of true knowledge recurs: what matters is not secular learning—which is useful, by the way—but instead union with God. The learned theologian wondered "What Euclid, what Marinus, what Ptolemy could have conceived of that? What Empedocles, Socrates, Aristotle, or Plato could have conceived of that with their logical methods and mathematical demonstrations?"[26]

According to Palamas, Plato's motto, "Let no-one ignorant of Geometry enter," ignored the fact that the true mathematician cannot separate the limit from what is limited and hence cannot gain knowledge of Creation. "The [anti-Hesychasts] cannot understand that God is simultaneously uncomprehended and comprehensible: uncomprehended in essence, but comprehensible by his creatures through His divine actions."[27]

The Orthodox Church officially awarded the victory to Palamas and supported the Hesychast movement against Barlaam and the humanists by a decision of the synod in 1341. Barlaam saw his anti-Hesychast ideas condemned by the synod, and he returned to Italy. Nikephoros Gregoras (see chapter 6) succeeded him as head of the anti-Hesychast party and found himself in opposition to the head of the Hesychasts, Gregory Palamas; he would even be imprisoned after the ultimate victory of Palamas. At Gregoras's death in 1360, his body was exposed to public view as if he were a criminal.

The church also succeeded in getting the emperors to choose the patriarch of Constantinople from among the followers of the Hesychast party. But more significant than official recognition was this movement's success in strongly marking not only Byzantine society but also Orthodoxy as a whole. It lay at the spiritual origin of the complicated relations between science and Russian society and also constituted the ideological basis of Slavic mysticism. Its consequences, right down to our day, are far from fully studied, but they have been well signaled by Russian intellectuals since the nineteenth century.[28]

This powerful movement that traversed the whole society did not, however, put a brake on the development of Byzantine humanism. This humanism embraced all the knowledge of the antiquity, especially philosophy, which notably included the philosophy of nature. Byzantium would increasingly discuss science in the fourteenth and fifteenth centuries. Nevertheless, it did curtail the eventual impulses toward subversive developments in the sciences; the Pletho phenomenon, named after a Byzantine scholar who returned to Hellenic religion, would remain an isolated exception (see chapter 9). It would make null and void any attempt at the union of churches, despite the keen efforts of several emperors. Byzantium would thus be condemned to Ottoman occupation, but the Orthodox Church would keep control over the Christian population of this region—right up until today.

CHAPTER EIGHT

Ancients versus Moderns

Byzantium and Persian, Latin, and Jewish Sciences

As we have seen, Byzantine scholars constantly taught, studied, and commentated on Greek science. However, the direct connection between ancient and Byzantine scholarship had been broken during the iconoclast period, which marked the entry of the Byzantine sciences into the Middle Ages. Stephen of Alexandria, the empire's œcumenic philosopher in the seventh century, was the last Byzantine scholar able to trace his academic lineage directly back to the ancient philosophers. His death symbolically marks the end of antiquity.

During the renaissance of scientific education in the ninth century, Byzantine scholars declared themselves to be the heirs of the ancient Greeks. Little by little, the term *Hellene*, which had had a negative connotation in the texts of the church fathers because it referred to pagan philosophers, became a positive notion for the erudite; henceforth, it referred to the ancient Greek scholars who built the foundation on which Byzantine science rested. Though sometimes contested, this ancient knowledge became a precious source of national pride.[1] Thus, Byzantines continued the ancient tradition of differentiating between Greeks and barbarians, a difference evidently founded on language, the vehicle of Hellenic culture. Throughout the Middle Ages and beyond, Byzantine scholars regarded the sciences of other peoples (ἔθνη) as inferior, even bordering on charlatanism.

Nevertheless, Byzantine scholars were soon taking an interest in certain aspects of the science of Islam, notably in the "technical" skill of Arab astronomy and its astronomical tables. The prime reason for this interest was that the planetary positions calculated following the Ptolemaic tradition (especially the *Handy Tables* based upon the commentaries of Theon of Alexandria) were, over time, presenting significant systematic discrepancies.[2]

We saw (in chapter 3) that the first influences of Arab science were detectable in Stephen the astrologer in 775, that during the same era Byzantine astronomers served Arab caliphs, and that at the start of the ninth century Leo the mathematician was invited to the court of al-Ma'mūn (chapter 4). These scientific encounters between the two worlds were not the only ones, and despite their hesitations, Byzantine savants increasingly eyed the Islamic side, if only for practical reasons: the Islamic tables were easier to use. Despite the fact that this science came from "unbelievers," using Islam's astronomical tables or its constants was a lesser evil for Byzantine savants. Indeed, the measurement of constants was founded on the observations so scorned by Byzantium, and the tables could be characterized as a simple technique not involving philosophical discussions on the world. During the whole Byzantine period, influences coming from Islam would be confined to practical astronomy and calculation, in particular the use of Arabic numerals (called *Indian*). The latter would never be adopted, though, because the tradition of using Greek figures was so strong.

After Stephen, the second Byzantine text that has come down to us in which we detect Arab or Muslim influence dates from the years following the first Byzantine humanism, when Greek science was well reestablished in education. In the margins of a beautiful ninth-century copy of the *Almagest* (manuscript *Vat. gr.* 1594) is found a *scholion* datable to the twelfth or thirteenth century, whose original text seems to have been written around 1032. The anonymous author compares the data of the tables of Ptolemy (*Almagest*, *Handy Tables*) with those of the *neōteroi*, the "new ones" or "moderns." But because he does not possess the tables of these *neōteroi* (the Arab astronomers of the time of al-Ma'mūn), he declares that he is using the tables of the astronomer Alim (ibn al-'Alam, d. 985).[3] Information on these tables is given in a short text titled *How to Make a Table according to Alim*, which gives the parameters of planets.[4] In addition, two horoscopes commissioned by a Byzantine in 1153 and 1162 were based on a *Summary of the Tables of Alim*, which shows that these tables enjoyed widespread diffusion and were used for at least a century.[5]

Shortly after the penetration by *neōteroi* science around 1072, *Methods of Calculation and Various Hypotheses* was composed in Constantinople by an anonymous author.[6] This Byzantine mentions that he has observed the solar eclipse of 20 May 1072—an exceptional fact, given a scientific culture that disdained observation, which on its own shows the Arab influence. The treatise gives instructions for the calculation of astronomical data and includes tables. The Arab sources of *Methods* are the commentary of ibn al-Muthannā (tenth century) on the *Zīj al-sinhind* of Muhammad ibn Mūsā al-Khwārizmī (first half of the ninth century) and the tables of Ahmad ibn Abdallah Habash al-Hāsib (middle of the ninth), notably his *Zīj al Dimashqī*. It is probable that *Methods* is also based on other Arab sources and specifically a tradition of Byzantine versions of Arab astronomy that have not survived. In fact, we notice a perfect Hellenization of the vocabulary for terms of Arabic origin (like the trigonometric functions other than the chords of Ptolemy), something that is absolutely astonishing for a text that could not have behind it a long tradition of previous texts.[7]

From this era also dates the sole Byzantine astrolabe that has been conserved, found today in the Civic Museum in Brescia. This instrument with obvious signs of Arab influence was constructed by (or for?) a certain Persian Sergios, who had the Byzantine titles of *protospatharios* and *hypatos*, in July 1062. The mold is engraved for Rhodes (36°) but there are also two plates, one for the Hellespont (40°) and another for Constantinople (41°, which is the latitude given this city by the Arabs).

These borrowings from Islam are due principally to the thirst of the Byzantines for astrology. Despite many (sometimes rather severe) condemnations of this science by the church, the emperors of the dynasty of the Komnenos were known for their taste for astrology. As we recall, Alexius I (1081–1118) had four astrologers at his court: two were Egyptians, one an Athenian, and the fourth the polymath Simeon Seth (see chapter 4). In his *Synopsis physikon* (*On natural things*), Seth, who had traveled in Egypt and Persia, mentions the Arab value of precession (1° in sixty-six years) and the list of stars of Apomasar (Abū Ma'shar).[8]

In the twelfth century, Arab astronomy seems to have been well known among the Byzantines, as witnessed by the manuscript *Vat. gr.* 1056, which includes a compilation of astronomical and astrological texts influenced by Arab astronomy.[9] This compilation contains:

- A treatise on the astrolabe titled *Various methods drawn from a Saracen book on the manner of taking with the astrolabe the horoscope and the twelve sites and knowing what is inscribed in the astrolabe* (The vocabulary is Greek, but one finds two Arab words.)[10]
- Three lists of stars: Seth's mentioned above, and two others that refer to the same sources, the *Zīj al-Hākimī* by ibn Yūnus (d. 1009) and one by Kūshyār ibn Labbān (c. 1009)
- Five astronomical tables, very incomplete
- The list of seven climates (zones of earthly latitudes), whose boundaries correspond more or less to those of al-Khwārizmī
- The horoscopes for the proclamation of Emperor Alexius I Komnenos on 1 April 1081, of Manuel I Komnenos on 13 March 1143, and the death of Emperor Alexander (not datable) (It may be that the positions of the

planets for these horoscopes had been calculated according to the *Zīj al-Hākimī*.)[11]

Apart from mathematical astronomy properly speaking, the Byzantines translated many purely astrological texts, notably the astrology of Abū Ma'shar, which exerted quite an influence.[12]

The documents mentioned here were undoubtedly only a portion of a much vaster corpus that is now lost. Some Byzantine astronomers of this period probably traveled to Islamic countries and knew Arabic and thus had access to assorted astronomic documentation in the Arabic language. One part of these texts was adapted into Greek, probably very early on, around the end of the eighth century. This would explain the Hellenization of the vocabulary that we find in later texts. The attested presence at the Komnenos court of Arab astrologers reinforces the hypothesis of a significant penetration of Islamic astronomy into Byzantium in the eleventh and twelfth centuries, despite the attitudes toward "barbarians" who were not even Christians. However, this Arab influence involved only the practical aspect of astronomy, the construction of tables, and especially the compilation of horoscopes. This leaves aside both the problematic of the world system, drawn directly from Ptolemy as the Byzantines' "own" astronomer, and cosmology properly speaking, which, because it intersected the domain of religion, could not be borrowed from Islam.

Persians against Ptolemy

The shock felt by Byzantine society after the conquest of Constantinople by the Christian Crusaders, which overthrew a political establishment that had endured for eight centuries, would lead this society to accept change as something that was henceforth probable. The Byzantines, so inclined toward tradition, were now forced to accept innovation, which brought about an important intellectual renaissance. The Byzantine Renaissance, also called the second Byzantine humanism (see chapter 5), not only was an intellectual movement of rediscovering and rereading the ancients but also consisted of discovering two other civilizations, viewed until then with enormous distrust, even contempt: Islam and the Latin West. This opening would entail a significant production of scientific literature, comparable in quantity to that at the end of antiquity. The field where the cultural openness was most visible was astronomy. The era of the Palaiologos saw such a rich production of astronomical works that it would be the most brilliant period for this science in Byzantium. Two currents competed against each other:

the restoration of Ptolemy (considered part of their own culture) and the Persian astronomy imported from Tabriz, in present-day Iran. To those were soon added Latin and Jewish influences.

Georges (or Gregory, his monastic name) Chioniades, who had studied in Constantinople after its reconquest by the Byzantines, was told that Persia was the ideal place to extend his knowledge of astronomy. We remember that this was an era when Constantinople lacked good experts in this science, since the *didaskaloi* of the imperial and patriarchal schools were stronger in natural philosophy. Very probably, not one of them was capable of performing the complicated calculations that eclipses required. Around 1300 Chioniades went to Trebizond, where the caravan left for Tabriz laden with goods brought from Europe by the ships from Genoa, to be traded for goods coming from all over Asia. There he managed to obtain the help of the emperor of Trebizond, Alexius II Komnenos (1297–1300), in order to travel to Persia. Once in Tabriz, having learned the language, he studied with the scholar Shams al-Dīn al-Bukhārī (b. 1254 in Bukhara). Shams taught Chioniades the astronomical learning of the School of Maragha, the famous observatory founded in 1259 by Hulagu Khan in his native city, in recognition of the good services that his astrologers had rendered him. The secrets of this science so useful to the Khans were reserved to their subjects, and so Chioniades had first to obtain the favor of Ghazan Khan in order to get permission to study there. Returning to Trebizond, he brought in his baggage Persian astronomical texts and a transcription of Bukhari's oral teaching. Chioniades must have gone to Persia several times, for he was named by the Holy Synod of the Church and the emperor of Byzantium Andronicus II Palaiologos (1282–1328) as bishop of Tabriz. This took place in a period when the Byzantines were trying to establish alliances with the Mongol princes; Andronicus II tried (without success) to marry his illegitimate daughter Irene to Ghazan Khan and, after the death of the latter in 1304, to his successor.

Thus, the Byzantine Chioniades became bishop of the Orthodox Church but saw no problem in studying astronomy—and astrology—with the Muslims. From his visits to Persia was born the Byzantine-Persian astronomical school. Current research attributes to Chioniades adaptations of various Arab or Persian astronomical tables and treatises on the construction of these tables, notably of the great Persian astronomer Nasīr al-Dīn al-Tūsī (1201–74), first director of Maragha's observatory, and on the theory of the astrolabe.[13] Chioniades also adapted a treatise on the astrolabe of Siams the Persian, which must be Shams Bukhārī's. The preface of this treatise, which seems to have been directly written in Greek, carries a dedication to Emperor Andronicus II, who was a great protector of astronomy. It is even possible that this text accompanied an astrolabe sent to the emperor.[14]

The dissemination into Byzantine scholarly milieux of this important corpus of Persian astronomy would require two generations of astronomers. In fact, after Chioniades, the corpus was found in the hands of a priest of Trebizond, Manuel, author of *Almanac of Trebizond for 1336*, which included astrological predictions, calculated from the Islamic tables *Zīj-i Īlkhānī* of al-Tūsī and *Zīj al'Alā'ī* of al-Fahhād (c. 1176).[15] The priest and astrologer Manuel taught Islamic astronomy to the doctor George Chrysokokkes, who went to Trebizond for this reason. He identified for his pupil the best treatise that Chioniades had adapted into Greek, which was none other than the *Zīj-i Īlkhānī*.[16] From the knowledge gained from both Manuel's teaching and the works in his possession, Chrysokokkes was able to compose around 1347 the *Persian Syntax of Astronomy*, which was widely circulated (more than fifty copies of the complete or partial text are extant).[17] The astronomical tables of the *Syntax* come from the *Zīj-i Īlkhānī* and the chapters on chronology from the *Zīj al-Sanjarī* of al-Khāzinī (c. 1135).

How can we explain this Persian success, even if delayed? After all, this science came from Persian infidels and, moreover, was linked to astrology, so often condemned by the Orthodox Church. In the first place, astrology, though banned in words, was well accepted in fact, as much by the emperors as by a number of educated clergy. As we have seen, Chioniades himself became a bishop, and Manuel of Trebizond was a priest. Ever since Saint Basil accepted that one might employ the stars as signs of the phenomena of nature (but not of human destinies), the debate had continued, not over whether to use planetary positions to make predictions but over the nature of these predictions. Metochites, a fervent enemy of astrology, thought that one could predict natural catastrophes with the help of the stars, and his pupil Gregoras went one step further in advancing that eclipses announced disasters.

Second, the *Persian Syntax* came three decades after a renewal of interest in Ptolemaic mathematical astronomy, provoked by the redaction of Metochites' *Elements of Astronomical Science* (see chapter 6), and fifteen years after the launch of eclipse calculations thanks to the polemic between Gregoras and Barlaam. In going farther than the formal study of the system of "their" astronomer Ptolemy, and in adopting the more complicated calculations, the Byzantines began to take account of the difficulties of applying tables prepared some twelve centuries earlier. Georges Lapithes, a Cypriot gentleman, wrote to Nikephoros Gregoras that "the Italians, with whom fate would have us cohabit, use Ptolemy very little for the two parts, I mean to say the theoretical and the practical parts, and rely more on the moderns. In fact, they are not content only with the Arab tables that begin with Mahomet, but use many others."[18] The tables of the *Syntax* of Chrysokokkes were adapted for the longitude of Cyprus around 1347, shortly after their redaction.[19] Even Nikephoros Gregoras, a fervent partisan of Ptolemy, suggested around 1332 that the tables of the greatest astronomer of all time had to be corrected.

The interest in non-Ptolemaic astronomies also stemmed from the difficulties of table construction. The tables of Ptolemy were truly out-of-date, but constructing new tables was a very difficult task. The best known were the *New Tables* of the monk Isaac Argyros (c. 1310–c. 1372), created at the beginning of the year 1367–68 and founded on Ptolemy. But Argyros, who was a committed anti-Hesychast, was condemned by the church, which threw an anathema on "Argyros, who was full of poison and had enriched himself too much on the false chatter of the Hellenes." The simplest solution to this technical problem was to translate the foreign tables, making the necessary conversions for the calendar and the longitude.[20]

Thus, despite the hesitation of the Orthodox world, the Persian tables ultimately enjoyed a great success. It should be noted, though, that this success cannot be explained by their better precision compared to Ptolemaic tables. Chrysokokkes, not possessing the commentary on the tables by *Zīj-i Īlkhānī*, committed errors (the meridian of origin, the alignment of celestial bodies called syzygies) that, when compounded, could sometimes give results less precise than those obtained by Ptolemaic tables.[21] So it is very probable that their success was due to how simple they were to use. Chrysokokkes presented the calculation of longitudes, latitudes, syzygies, and eclipses—everything necessary to establish astronomical ephemerides, astrological *thematia*, and the date of Easter. His goal was practical, and anything that was not necessary to this kind of calculation, such as trigonometric functions, disappeared. Thus, despite efforts to update Ptolemaic astronomy, the Persian tables became prevalent. The reluctance to use them in educated ecclesiastical circles quickly dissipated, and the tables were consecrated by the *didaskalos* of *didaskaloi* of the Patriarchal School, the great scholar Theodore Meliteniotes (c. 1320–93). (It is unknown whether this title was equivalent to rector of the Patriarchal School or to spokesman of the emperor at this school.) In the image of the Trinity, Meliteniotes composed three books in three parts, the *Holy Tribiblos*, the *Astronomical Tribiblos*, and the *Triple Exegesis of the Trinity*. The three parts of the *Astronomical Tribiblos*, composed around 1352–68, deal with arithmetic and the astrolabe, with Ptolemaic astronomy (*Almagest* and

Handy Tables), and Persian astronomy. Possessing an open mind, Meliteniotes chose the same examples for the Ptolemaic and Persian parts of the *Tribiblos*, which gives the reader the possibility of comparing the efficacy of both methods. Given the errors already cited, the comparison is not always favorable to Persian astronomy. Despite that, whereas the *Tribiblos* was not widely circulated (ten surviving manuscripts, of which only two are the complete text), its third part, titled *Teaching the Persian Tables* and transmitted (more or less reworked) under the names of either Isaac Argyros or George Chrysokokkes, has survived wholly or in part in thirty manuscripts. Perhaps Byzantine society was discovering a taste for exoticism.[22]

After Meliteniotes' effort at synthesis, Byzantine astronomers abandoned their scruples about using the science of the infidels and went about comparing the performance of the Ptolemaic and Persian tables, especially when it came to eclipses. A list of solar and lunar eclipse predictions from 1376 to 1408, dated about 1364–75, compares the two methods in detail.[23] Sometimes the syntheses were subject to experimentation, as with the notary of the Great Church and future bishop of Selymbria, John Chortasmenos, for the solar eclipse of 15 April 1409. He used the Ptolemaic tables of Isaac Argyros, but for the parallaxes he preferred the Persian table, which is extremely simplified.[24] It is by pure chance that the result coincided almost exactly with modern calculations.

Contacts with the West and the Coming of Latin and Jewish Science

The Orthodox Church, well before the schism that officially took place in 1054, considered the pontifical church in Rome as heretical and viewed the "Latins" with distrust. The feelings were reciprocal; this animosity, nourished by the Eastern schism (according to the West) or by the Latin schism (according to the East), culminated in the seizure of Constantinople in 1204.

Despite this quasi-permanent confrontation, efforts at reunification of the Orthodox and Roman Catholic churches never truly ceased on one side or the other. Manuel I Komnenos (r. 1143–80) had the ambition of restoring a Roman Empire that could be achieved only by means of church union. Profiting from the bad relations between the Italians and the Normans dominating their lands (Sicily), he conducted an offensive policy in Italy and concluded an ephemeral alliance with the papacy. The wars of the Byzantines against the Kingdom of Sicily led to negotiations and contacts that had the unexpected effect of the first direct medieval translations into Latin of Greek scientific texts.

Henricus Aristippus (d. c. 1162), archdeacon of Calabria, was a connoisseur of Greek and a lover of the sciences. He spent 1158–60 in Constantinople as envoy from the Kingdom of Sicily. On this occasion, the emperor Manuel I Komnenos offered him a copy of the *Almagest*, which Henricus brought back to Sicily. This manuscript was later translated into Latin by one of his students. Aristippus himself translated the *Meteorologica* of Aristotle and the *Pneumatics* of Heron of Alexandria, as well as Greek works that did not involve science. It is to the circle of "Sicilian" translators that we owe the translations into Latin of the works of Aristotle, Heron of Alexandria, Ptolemy, Euclid, and Proclus.[25]

These translations did not have a major impact in the West at the time, no doubt because of the lack of a community of scholars capable of exploiting them in the south of Italy. Moreover, the transmission went in only one direction. On the Eastern side, Byzantine savants had nothing to expect from the Latin savants, looking down on them as uneducated for not knowing the works of Aristotle and Ptolemy.

The rapid development of the study of Greek science in the West, along with the contacts of Byzantines with this learning via the current Latin domination of a large part of the territory formerly controlled by their empire, would gradually transform this contempt into interest. This change began to manifest itself after the retaking of Constantinople by the Byzantines; it was strengthened by reactions to the Hesychast debate, which had demoralized many Byzantine scholars. They now turned toward the Catholic West for the recognition that they thought they could not find at home.

The place where Byzantine and Latin cultures, Orthodoxy and Catholicism, were longest in contact was Cyprus. Conquered by Richard the Lionheart in 1191, it passed to the Lusignans in 1192, and they kept it until 1489, when Venice took possession. In the second quarter of the fourteenth century, the Greek gentleman George Lapithes gathered a circle of scholars and lovers of astronomy and other subjects. This circle possessed treatises in Latin of Arab inspiration that its members translated into Greek. Thus, around 1337–40, the *Toledan Tables* were adapted, probably by Georges Lapithes himself.[26] In the same manuscript is found a treatise on the construction and use of the astrolabe, which is also based on Latin sources of Arab inspiration, notably the treatises of Messahalla and Maslama.[27]

Within a few decades, the Byzantines were using the Latin science that they had so long disdained. Demetrius Chrysoloras (d. c. 1417) was well known in Renaissance Europe, as was his brother Manuel, confidant of the emperor-philosopher Manuel II Palaiologos (r. 1391–1425). After the polemic over Hesychasm,

Byzantines embarked on a discussion of the theses of the thirteenth-century Italian priest-philosopher Thomas Aquinas. Demetrius authored an anti-Thomist treatise but in a conciliatory tone. This theologian knew Euclid and Ptolemy, and he taught science to scholars such as John Chortasmenos. The end of the fourteenth century was an era of intense diplomatic activity aimed at Italy, whose help was solicited against the Ottoman advance. Greek scholars went to Italy on diplomatic missions and often brought back in their luggage scientific manuscripts in Latin. Thus, around 1380 Chrysoloras would translate into Greek and adapt for Constantinople the famous Alphonsine tables, composed in honor of King Alphonse X of Castile more than a hundred years earlier, in 1252, but still widely used in Italy. This adaptation was one of the rare examples of penetration into the Orthodox world of purely Latin science. The contacts by the Orthodox with the West, whether through diplomatic missions or via the Latin possessions in the eastern Mediterranean, would lead to the importation of Latin-composed science of Arab or Jewish provenance, as was the case of adaptations done by George Lapithes.

Marc Eugenicos (1393–1445) was much more engaged ideologically than Chrysoloras. A supporter of Palamas, he was leader of the anti-unionists. Apart from his religious pursuits, he studied science with Pletho, John Chortasmenos, and Manuel Chrysokokkes. In 1418 he became a monk, and in 1437, supported by Emperor John VIII Palaiologos, he was proclaimed bishop of Ephesus so that he could participate in the Council of Ferrara/Florence in 1438–39 to discuss church union. He was the only Orthodox representative to refuse to sign the accord in favor of union, and he subsequently worked to abort it. His fundamental disagreement concerned the problem of *filioque*. In his discussions with the Catholics, in which he argued on the basis of Basil's *Hexaemeron* among other works, he refuted the necessity of purgatory (which does not exist in Orthodox theology) for the simple reason that for the souls of the dead there can exist no time of waiting for the resurrection, since they are outside Creation, in a nonplace where the notion of time does not exist. Glorified by the people of Constantinople, who had strong anti-unionist feelings, he was among those (such as his student and first patriarch under the Ottoman domination, Gennady Scholarios) who saw the advance of the Turks as a lesser evil than submission of the Eastern Church to the papacy. During his patriarchate, Scholarios proclaimed him a saint of Orthodoxy.

Marc, a fervent partisan of the independence of the Eastern Church in the face of the Latin heretics and a defender of Hesychasm, remained a great lover of secular learning and did not hesitate to use his stay in Italy to acquire astronomical manuscripts that were unknown in Byzantium. These Latin manuscripts, however, were not from the Latin tradition—but Jewish. Marc adapted around 1448 the *Cyclical Tables* of Jacob ben David Yom-Tob of Perpignan, written around 1361. Marc called it the *Method concerning the New Handy Tables Composed in Italy.*[28]

Over time, Byzantine Orthodoxy showed more openness to the secular learning of Muslims and Jews than to that of their fellow Christians, the Catholics. While the fourteenth century marked the acceptance of Persian science, the fifteenth was that of Jewish science. An important factor in this encounter was the Karaite community of Jews.

Karaism has often been compared with Protestantism. The Karaites recognized only the Hebrew Bible and rejected the postbiblical tradition carried in the Talmud, as well as the rabbinic tradition. In the fourteenth and fifteenth centuries, the communities of Karaites around the Mediterranean were in decline; they turned in on themselves and, as a result, became a significant network of exchanges. Karaite travelers were always welcomed and found refuge in their communities, where they exchanged experience and learning. At this time, Constantinople possessed a significant Karaite community that developed into an intellectual center, especially in mathematical sciences. The scientific tradition of this community survived the Ottoman conquest, and its best-known mathematician was Caleb Afendopoulo, who was born after the conquest (1455–c. 1509).

During the thirteenth and fourteenth centuries, the Karaite community of Languedoc in French Provence developed a school of mathematics and medicine that played a major role in the spread of science in Europe. The mathematician Immanuel ben Jacob Bonfils of Tarascon wrote several books, one of which had an unexpected success. It was called *Kanfe nesharim* (the wings of eagles, an allusion to Exodus), which he wrote around 1365 and which was known by the title *Sepher shesh kenafayim* (book with six wings) since its tables were divided into six groups or wings (an allusion to Isaiah 6:2); they were astronomical tables with instructions on how to use them. Thanks to this book, Bonfils became known under the sobriquet "master of wings."[29] Several *scholia* on these tables, calculated for the longitude of Tarascon, were written by Karaites, who gave information about making the calculation for Constantinople and Crimea. These *scholia* were translated into Latin in 1406 by Pico della Mirandola, as well as into Russian.

Michael Chrysokokkes was a grand church notary, and hence was close to the patriarch of Constantinople around the second half of the fifteenth century. Thanks to the Karaite community of Constantinople, it is probable that Michael knew Bonfils's original text in Hebrew, which he translated and adapted for Constantinople around 1434. This adaptation also had an unexpected success: more than fifteen copies survive today, as well as *scholia* and complements, such as one by Stouditis of Damascus in the sixteenth century. Since the prime function of these tables was to calculate eclipses, we can understand their success in the Byzantine context of the day.

Unionists, Anti-Unionists, and the Sciences of the "Moderns"

One might think that science went into decline after the victory of the Hesychast party within the Orthodox Church. If I have stressed the science of astronomy, it is because its proliferation belies this supposition. Moreover, it is an astonishing fact that almost all Byzantine astronomers of the last hundred years of the empire were church dignitaries. Some of them were the sworn enemies of union, while others were affiliated with the Hesychast movement. Thus, one cannot maintain that science was developed only by the partisans of union and the enemies of Palamism, or that the latter were more open to Latin science. The senior church dignitaries of this period, whether unionists, anti-unionists, or Hesychasts, all seem to have been knowledgeable and enthusiastic about mathematical astronomy and to have competed with one another to produce the best tables and most precise calculations of eclipses.

But why this ferment? We have noticed that astrology was never clearly excluded from the interests of the Orthodox Byzantines. And the calculation of Easter was always a major preoccupation of the church. Nor should we underestimate the pure and fashionable skill of eclipse calculation, which required that a good scholar had to be an expert in these matters.

The involvement of the high clergy in astronomy would not have been possible without the higher education in this science at the Patriarchal School that had started in the middle of the fourteenth century. The *didaskalos* of *didaskaloi*, Theodore Meliteniotes, an expert in both Ptolemaic and Persian astronomy, was probably at the origin of this advanced education. Until Meliteniotes, owing to the direct implication of astrology in astronomy, the church had hesitated to teach it in its schools. But the *Tribiblos* of Meliteniotes did not promote astrology; it took up the tradition of mathematical textbooks and its pedagogic intentions are evident and of high quality.[30] This manual's content placed astronomy, whether in the Ptolemaic tradition or not, within the scope of Orthodox dignitaries. Meliteniotes was thus the first professor in a series of teachers that led from Demetrios Chrysoloras, John Chortasmenos, and Marc Eugenicos to Bessarion. The influence of this school extended for a short time as far away as Ukraine, via the Unionist Isidor of Kiev (or of Russia, c. 1385–1463), a Greek from Thessalonica who had studied in Constantinople and was appointed metropolitan of Kiev, Moscow and all Russ, which means he was head of the Russian Orthodox Church. Isidor wrote texts of eclipse calculations and several on astronomical treatises.[31]

The Hesychast movement and Orthodox opposition to the union with Catholicism, which would facilitate the Ottoman conquest, did not have negative implications for education in the mathematical sciences at the Patriarchal School. A decade after the official victory (in 1341) of Hesychasm, the spiritual movement that had disdained secular learning in general and Greek philosophy in particular—the same group that had cast anathema on the leader of anti-Hesychasm, Barlaam—would establish in its most prestigious school (the Patriarchal School) the most advanced teaching of astronomy. Its students, whether partisans of Hesychasm or the union of the churches, all showed unexpected interest in mathematical astronomy and were open to non-Orthodox knowledge, whether coming from Muslims, Jews, or Catholics. Despite the Hesychast reaction, Byzantium in the fifteenth century participated in the scientific Renaissance, thereby continuing the humanist tradition of the preceding century. But soon the irresistible Ottoman advance and its conquering of Constantinople would put an abrupt stop to this movement. Many Byzantine scholars fled to Italy and other countries of western Europe. Henceforth, they would participate directly in the European Renaissance.

CHAPTER NINE

The Fall of the Empire
and the Exodus to Italy

In its final effort to survive the irresistible Ottoman advance, the feeble and impoverished Byzantine Empire turned to the West. The Byzantine emperors sent to Italy the finest flower of Byzantine intellectuals to take part in negotiations about uniting Orthodox and Catholic churches. The endless discussions about matters of dogma brought together Greek scholars and those of Latin Europe. During their stay in Italy, Byzantine envoys had the opportunity to diffuse Greek culture, language, and science, and in doing so, they contributed to the birth of the Italian Renaissance.

The Pagan Calendar of Gemistus Pletho

In the last Byzantine decades, the most remarkable figure in philosophy and science was Georgios Gemistus Pletho. Born in Constantinople around 1355–60, he received an excellent education that included the *trivium* and the *quadrivium*. He also acquired solid knowledge of astronomy and mathematics, probably from a commentator on Diophantus named Demetrius Cydones (1324–28), theologian, statesman, fervent partisan of the union, and a convert to Catholicism. But what most distinguished him was his study with the Jew Elisha at the Ottoman court of Adrianopolis.

When Sultan Murad I transferred the Ottoman court from Bursa to Adrianopolis, a community of Jewish scholars formed there, whose learning was of great interest to the Greeks. As formerly when some Byzantines went to teach or study in the Arab world in the midst of clashes between Islam and the Byzantine world, so now during the wars between Ottomans and Byzantines some scholars continued to live among the Turks. Pletho went to Bursa to learn from the scholar Elisha about Averroes' interpretation of Aristotle. Averroes (1126–98), who believed that there is no conflict between science and religion because the former is based on study and the latter on faith, had developed a theory of motion different from Aristotle's by introducing the notion of inertia in celestial movement. Averroes had also related force to the change of the kinetic condition of a body, developed a new theory of vision, and written a monumental medical encyclopedia. Although his work had an important influence on Catholic scholars, notably on Thomas Aquinas, he did not have much affect on Byzantine science. Averroes' work did, however, have a notable influence on Jewish philosophers. Elisha was a Jew and probably a doctor at the Ottoman court; he knew Aristotle's work, but he was also a partisan of the "Illuminationist" current of the Persian Suhrawardī (1155–91), who admired Zoroaster and Plato and aspired to resurrect the glory of ancient Persia. Inspired by the Neoplatonists, Suhrawardī founded the Persian school of Illumination, which believed that light produces intellect and that souls are inspired by a metaphysical being.[1] Thus, apart from Averroes' commentaries on Aristotle, Elisha taught Pletho Illuminationism, which profoundly marked the young Greek scholar.

Returning to Constantinople, Pletho taught science and philosophy until 1410, when he was sent, perhaps exiled, by Emperor Manuel II to Mistra in the Peloponnese, where he would spend the remainder of his life. At Mistra he developed Neoplatonist and pagan ideas that he expressed in his *Book of Laws*, in which he promoted a plan for the re-Hellenization of the Peloponnese. The greatest Byzantine philosopher of his time advanced the idea that the ruin of the empire was due to the Christian religion.

In his later years, he joined the Greek delegation at the council that met at Ferrara and Florence in 1438–39 to discuss the union of the Catholic and Orthodox churches. More than eighty years old at the time, he aroused general admiration among Italian scholars, especially for his expertise on Plato. It was during his stay in Italy that he changed his name from Gemistus to its synonym Plethon to resemble Platon (Plato). Within a very large Greek delegation were his students Marc Eugenicos and Bessarion, as well as Pletho's future enemy, Scholarios. Pletho as a secular adviser did not participate in all the council's sessions, so he spent part of his time in Florence propagating Plato's ideas, which, compared to those of Aristotle, were almost unknown to the Italians. One consequence of his propaganda was the founding of the Academia Platonica in this city by Cosimo dei Medici (1389–1464). Bessarion, one of the great scholars of the Italian Renaissance, wrote that Pletho "was really an image of philosophy and of any kind of science, not only as concerns writing, but in anything concerning the courses of the stars, harmonic relations, geometric proportions, and arithmetic mediations."[2]

Pletho was a major exception in the Byzantine world: he had become a pagan, as he openly proclaimed in his *Book of Laws*. His philosophy of nature, and even his astronomy, reflected his ideas of returning to Hellenism. He saw Pythagoras and Plato as the heirs of Zoroaster and dreamed of a reconquest of Greece starting from Mistra, which he identified with Sparta on account of their geographical proximity. His astronomical work is of great interest, because Pletho proposed a non-Christian calendar linked to his ideal of returning to Hellenic sources. He titled a book he wrote around 1433 *By George Gemistus the philosopher, A method to find the conjunctions of the Sun and Moon and the full moons and the positions of the stars, According to the tables that he himself has established.*[3]

Pletho knew astronomy well: he understood Ptolemy, and the astronomies of the Arabs (the tradition of al-Battanî), Persians (tradition of al-Tusī), and Jews (tradition of Bonfils of Tarascon). Despite some weakness in mathematics, he demonstrated erudition.[4] In his book on the sun and the moon, as well as providing the astronomical tables that he created, he offered a short introduction in which he defined astronomical terms, explained briefly the calculation of syzygies (alignment of sun-earth-moon), and discussed the true positions of the sun, the moon, and the five planets. Unlike contemporary Byzantine astronomers, Pletho did not present all this in order to be able to proceed to the calculation of eclipses or horoscopes. Instead, his goal was to establish a new calendar, very different from the one used by Christian countries. In one of the surviving fragments of the *Book of Laws*, we find a nontechnical description of a new calendar that even has its own particular festivals.

Christians used a solar calendar based on the solar year, not on the revolution of the moon; the only time the moon was involved was for the calculation of Easter. Pletho wanted to return to the idea of the ancient Greeks, who had fixed the months according to the moon and the years according to the sun. His goal was to establish a lunar-solar calendar. The Muslims used a strictly lunar calendar, but other civilizations, including the Chinese and the Jews, used such a mixed calendar. The Chinese had lunar months and occasionally a short intercalary month to catch up the year. We might suppose that Pletho had no information concerning the Chinese calendar, but we do know that he knew Hebraic astronomy and made use of it.

According to Pletho's proposed calendar, the day begins at midnight, the month on the first midnight that follows the new moon, and the year on the first month that follows the winter solstice. Since the lunar cycle is a little more than 29.5 days, a month of 30 days succeeds a month of 29 days, and sometimes one finds two months of 30 days in succession. According to this system, the year includes either twelve months or thirteen in a sequence that returns every nineteen years. Its length varies, comprising either 354 or 355 days in the case of the year of twelve months or 383 or 384 days in the case of the year of thirteen months. Pletho, like the ancients, distinguishes between the solar year (*eniautos*, one revolution of the sun around the zodiac) and the calendar year (*etos*). He divides the month into seven unequal parts linked to the phases of the moon: day 1, days 2–8, days 9–14, day 15, days 16–22, days 23–28, and day 29 (or 29–30). Pletho's calendar was a challenge to Christianity, but because of its limited diffusion, it did not engender significant debate. Even commentaries by humanist scholars were negative. Among the critics was Theodore of Gaza (c. 1400–1475) a Greek scholar who took part in the Council of Ferrara-Florence in favor of union. He immigrated to Italy, translated Aristotle and other Greeks into Latin, and became rector of the University of Ferrara, where he founded an academy to match the Platonist one of Florence. Like many of his contemporaries, he knew of Pletho's calendar proposal through his *Book of Laws* and not through his astronomical work itself; Theodore did not hide his mockery of the great scholar and teased him for not giving names to the months but simply numbering them.[5]

Mehmed the Conqueror and the Greek Scholars

On 29 May 1453, after a two-month siege, Sultan Mehmed II the Conqueror entered on horseback the Church of Saint Sophia in Constantinople, the supreme emblem of Orthodoxy, while the queen city was being sacked. In addition to being a warrior, Mehmed was a subtle politician. He sought to alienate the Greeks from the Westerners in order to reduce the chance of a crusade by Christian countries against the Turks. For this purpose, he supported the enemies of church union, and Scholarios was perhaps the most notable of them.

George Kourtesios Scholarios (c. 1400–1473), adviser in theological affairs to John VIII Palaiologos, also participated in the Council of Ferrara-Florence, along with most of the important Byzantine scholars, in order to promote the union of churches. An Aristotelian, he found himself faced with the Platonist Pletho. Returning to Constantinople after the council, Scholarios, as some other partisans of union did, changed camps and became a fervent enemy of the Catholic Church. Soon after the conquest, Mehmed, learning that Scholarios had been made prisoner during the sack of Constantinople, liberated him and placed him at the head of the first millet of the Ottoman Empire (see chapter 10). After raising him rapidly through all the ecclesiastical grades, the sultan appointed him Patriarch Gennady (Gennadios) II on 6 January 1454.

In the spirit of "the Turks are better than the Latins," Scholarios maintained good relations with the sultan and inaugurated the Ottoman period of the Orthodox Church. Having control over the Orthodox Christian millet, he tried to purge it of nefarious influences, whether Latinizing or (worse) Hellenizing. In 1453, shortly after the death of Pletho, he attacked him for having studied ancient writers not in order to perfect the Greek language, as suited a good Christian, but for their ideas. Yet Scholarios himself had also studied the ancients for their ideas. He admired Aristotle and was perfectly familiar with secular literature. But he could not accept Pletho's idea that Christianity was responsible for the ruin of the Roman Empire. Nor obviously could he accept Pletho's militant Zoroastrianism. Pletho, he declared, "forgets that Zoroaster, celebrated among the Persians for astronomy, was not known for anything else, and was led astray by it, and moreover proved the *predictions* conforming to his *horoscope*. In fact, he was the son of Ninos, who married his own mother, Semiramis."[6] Accordingly, Scholarios made an uncommon decision for the Eastern Church: to burn publicly Pletho's *Book of Laws* as Hellenizing (idolatrous) and satanic.

The hold of the most fervent anti-unionists over the Patriarchate of Constantinople, as well as the inevitable withdrawal of the Orthodox Church after the Ottoman conquest, distanced the church from secular learning. Unlike Pletho, who believed that Christianity was responsible for the defeat, the most zealous anti-unionists believed that the Greeks were paying for their sins, including their connivance with the West and their involvement in Hellenistic learning. The wind favorable to the sciences that had blown during the Palaiologan dynasty abruptly changed direction.

Greek Schools and Manuscripts in Italy

One of the first acts of Gennadios Scholarios was to reopen the Patriarchal School in 1454, naming as director Matthew Camariotes, already involved in this school before the Ottoman conquest. But this new institution, like the new patriarchate, was installed in a small church rather than in the majestic Saint Sophia, which had become a mosque. It was now only a shadow of what it had been before the conquest. The new curriculum, although it probably included philosophy, left no trace of the teaching of science. Theology was henceforward absolutely dominant. Apart from the organizational and financial problems, it lacked the essential intellectual milieu for the development of any scientific study worthy of the name.[7] Indeed, after the Ottoman conquest, most of the Byzantine scholars, carrying away the manuscripts they had managed to save, fled to Italy on Genoese ships. Others with more foresight had already established themselves there. In addition, certain Greeks–for example, George of Trebizond (1395–1486), born in Venetian Crete—immigrated to Italy because they originated from Venetian possessions. George settled around 1430 in Italy, where he became known for his erudition on Aristotle and his anti-Platonism. His work would place him within the Italian scientific tradition and not the Greek Orthodox one.

Yet Mehmed the Conqueror, after a three-day sacking of the most splendid city of the Orient, treated the few remaining Greek scholars rather well. For example, the mathematician and philosopher George Amiroutzes had been a dignitary under the last emperor of Trebizond. When the Ottomans occupied the town in 1461, they captured him. We find him again that summer teaching geography to Mehmed; later, he translated for the sultan the geography of Ptolemy and drew a map of the world. Becoming the sultan's protégé, he even acquired enough influence to make and unmake patriarchs.

The whole Byzantine education system ceased to exist at the same time as the empire: no more schools, no more libraries, almost no scholars. The Greek manuscripts kept in the library of Topkapi Palace could not be consulted by the few remaining scholars, except for protégés of the sultan such as Amiroutzes. They were sold off in the course of the coming centuries.[8] Men, manuscripts, and education would all be transferred to Italy. There, Byzantine scholars found the support of Bessarion, whose fame was already well established among the Catholics.

Bessarion, whose secular name was Basil, was born around 1395 in Trebizond; he studied with George Chrysokokkes and, around 1431–35, with Gemistus Pletho. Upon the ephemeral union of churches at the Council of Ferrara in 1439, it was he who delivered the proclamation of union in Greek, in the presence of Pope Eugene IV and Emperor John VIII Palaiologos. Named cardinal by the pope, he quit Constantinople definitively in 1441 because of the fanatically anti-unionist ambience that prevailed there. He quickly acquired the reputation of a great scholar. His library, which he had brought from Byzantium and which he expanded in Italy by purchasing Byzantine manuscripts, comprised more than nine hundred codices. He offered it to Venice in 1468, where it became the core of the Biblioteca Marciana. Bessarion's collection was not the only one to enrich the Italian libraries. Thousands of Byzantine manuscripts were sold or donated to the West, contributing to the European Renaissance. A famous example in science is the manuscripts of the Vatican library consulted by Copernicus around 1500, which inspired him to develop his planetary theory.[9]

In Italy, a circle formed around Bessarion of Greek literati who had fled the

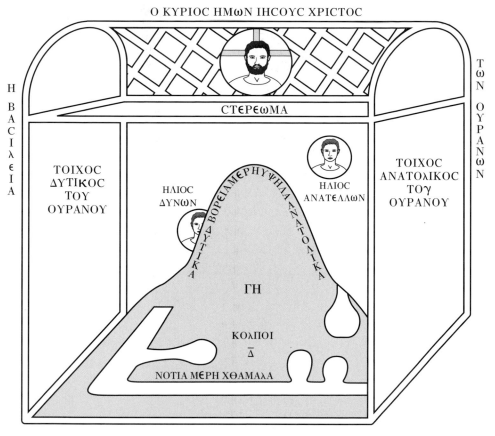

Figure 1. Cosmas Indicopleustes' conception of the world in two kinds of space.

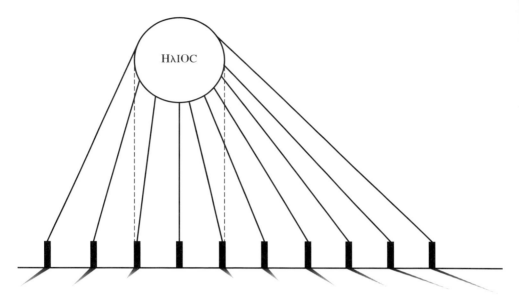

Figure 2. Cosmas Indicopleustes' conception of the sun's shadow on a flat Earth.

Figure 3. Saints Paraskeve, Gregory the Theologian, John Chrysostom, and Basil the Great, early fifteenth century. Tretyakov Gallery, Moscow. HIP / Art Resource, NY.

Figure 4. Emperor Alexander VII (912–913), the briefly reigning brother of Leo VI the Wise. Mosaic from the tympanum in the narthex of Chora Church, Istanbul. Byzantine, fourteenth century. Erich Lessing / Art Resource, NY.

Figure 5. Russia officially becomes a Christian nation with the conversion of Vladimir, Prince of Kiev, in 988. An Orthodox mass in Byzantium witnessed by Vladimir's envoys; the envoys report to the Prince. From the Radziwill Chronicle, page 59 v., 612 miniatures, late fifteenth century. Academy of Science, St. Petersburg. Erich Lessing / Art Resource, NY.

Figure 6. The heavenly ladder. Illustration of an instruction to monks by abbot Saint John Klimax (Climacus or Saint John of the Ladder, c. 570–650 CE). Good monks climb steadily heavenward toward perfection; bad monks are dragged to hell by black devils. Icon, late twelfth century. St. Catherine Monastery, Mount Sinai, Sinai Desert. Erich Lessing / Art Resource, NY.

Figure 7. Model of the only Byzantine astrolabe entirely preserved, of Persian inspiration, constructed in 1062. The original is kept at the Civici Musei d'Arte e di Storia in Brescia. Courtesy of the Science Center and Technology Museum (NOESIS), Thessalonica.

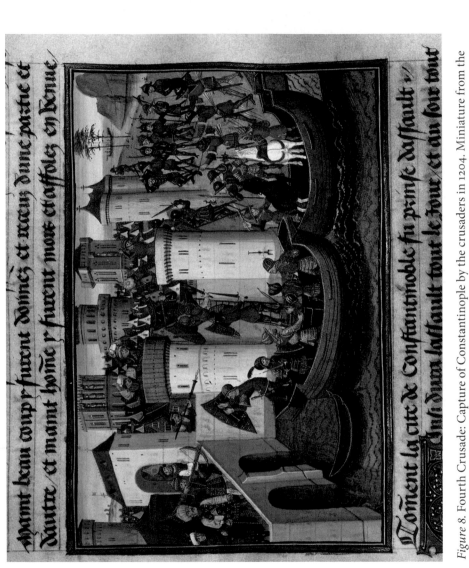

Figure 8. Fourth Crusade: Capture of Constantinople by the crusaders in 1204. Miniature from the *Chronique abrégée* by David Aubert, fifteenth century. Bibliothèque de l'Arsenal, Paris. Scala/White Images / Art Resource, NY.

Figure 9. The great *logothetēs* and astronomer Theodore Metochites (1270–1332), who restored the fifth-century Chora Church. Mosaic in the narthex of Chora Church, Istanbul. Byzantine, fourteenth century. Erich Lessing / Art Resource, NY.

Figure 10. The sky. From the Revelation of St. John: "and the heaven departed as a scroll when it is rolled together." Pareccelesion of Chora Church, Istanbul. Byzantine, fourteenth century. Werner Forman / Art Resource, NY.

Figure 11. Portrait of Cardinal Bessarion (1403–1473), sixteenth century. Accademia, Venice. Cameraphoto Arte, Venice / Art Resource, NY.

Figure 12. Portrait of Cyril Lucaris (1572–1638), patriarch of Constantinople. 1632. Bibliothèque Publique et Universitaire, Geneva. Photo: G. Dagli Orti. © DeA Picture Library / Art Resource, NY.

Figure 13. Chrysanthos Notaras (1663–1731), patriarch of Jerusalem, engraving, *Eisagogi eis ta geografika kai sfairika*, Venice, 1718. Courtesy of the Institute for Neohellenic Research, National Hellenic Research Foundation, Athens.

Figure 14. Balanos Vassilopoulos by A. Zuliani, engraving, *Odos Mathematikes*,
Venice, 1749. Courtesy of the Institute for Neohellenic Research, National Hellenic
Research Foundation, Athens.

Figure 15. Eugenios Voulgaris (1716–1806), engraving, *Ekatontaetiris ton tou Christou Peniton*, 1865. Courtesy of the Institute for Neohellenic Research, National Hellenic Research Foundation, Athens.

Ὄψει ὄψιν ἔχεις κραδίης σὲ φέροντος ἔσωθεν
Σκαρλάτε Στέρξα κύδιμε Θεοτόκη!

Figure 16. Nikephoros Theotokis (1731–1800), engraving, *Stoicheia geografias*, Vienna, 1804. Courtesy of the Institute for Neohellenic Research, National Hellenic Research Foundation, Athens.

Ottoman conquest. Religious differences mattered little. Bessarion had become a Catholic, whereas most of the Greek scholars remained Orthodox. Their common purpose was to persuade the pope and the Christian states to organize a crusade against the Ottoman Turks by reconquering Constantinople, and this united them more than religion divided them. Moreover, the Orthodox Greeks who settled in Italy did not include the most fanatic anti-unionists. The bizarre Bessarion, though now a Catholic cardinal, continued to wear the Orthodox beard, which did not always please his colleagues in the Papal See conclave. His appearance may have something to do with the fact that he failed twice (narrowly) to be elected pope.

Venice, because of its wealth, its ancient relations with Byzantium, and its Greek possessions (it kept Crete and the Ionian Islands, although it would soon lose the Peloponnese and the Island of Euboea to the Ottomans) was the preferred destination of Byzantine scholars. After the fall of Constantinople, this city became the center of Greek studies. This reputation resulted in part from the Greek printing press of Aldus Manutius (1450–1514), who, with his Cretan collaborator Marcus Mousouros, benefited from the wealth of Bessarion's library to print an exceptional series of ancient Greek texts. Manutius, protégé of the Renaissance philosopher Pico della Mirandola (1463–94), published the first Greek book in Venice in 1495. In the same period, another Cretan, Zachary Kalliergis, founded a second Greek printing press in Venice in which only Greeks worked. After the election of the humanist pope Leo X, Kalliergis founded a Greek press in Rome; its first book appeared in 1515.[10]

After the failure of the union between Catholic and Orthodox communities, the Roman Catholic Church went on the offensive to win over Orthodox communities, having some success in Venetian-dominated territories, especially the Aegean Islands. But it was not easy to influence communities in the Ottoman world, as they were under the protection of the sultan, who wanted to retain control over a church that was independent of Rome. In addition, the upper hierarchy of the Orthodox Church became fanatically anti-Western after the fall of Byzantium. For some, the Ottomans were preferable to Catholic Christians, because there was less danger that an Orthodox would convert to Islam than to Catholicism.

The creation of a college under the control of the papacy fit perfectly into the policy of expansion of the Catholic Church in the East. Colleges under religious control were in vogue in the sixteenth century. They prepared pupils who wanted to have an ecclesiastical career, and they also offered general culture with a solid grounding in theology to those who intended to remain secular. So in

1514 Pope Leo X founded the Greek College of Rome and called on the Greek humanist Ianos Laskaris (c. 1445–1534) to organize it. Around 1530, Laskaris also helped to found the Collège de France in Paris, where he taught ancient Greek. The Greek College of Rome did not survive Pope Leo X, who died in 1521. But fifty years later, in 1576, another reforming pope, Gregory XIII, who radically revised the Christian calendar, founded the Greek College of Saint Athanasius in Rome (which still functions today) and placed it under Dominican control. The college accepted all young Greeks irrespective of their religious affiliation but trained them in Catholic theology. Many partisans of the union graduated from it, as, for example, Leo Allatius (1586–1669) from the island of Chios, student and afterward professor at this college. Allatius, a physician, in one of his books, *De Graecorum hodie quorundam opinationibus* (1645), gave a scientific discussion of the mystery of vampires—and the opinions of the Greeks on this subject.

The same policy as at the College of Rome, of accepting young Greeks of both Catholic and Orthodox faiths, was followed by the San Giovanni College of Padua, which was founded in 1623. Forty years earlier, the rich Cretan Catholic priest Iosaffat Palaiokapas left in his will to the city of Venice a scholarship endowment for eighty-four youths from Crete and the Ionian Islands to study at the College of Saint Athanasius and the University of Padua. But when in 1623 the pope substituted the Jesuits for the Dominicans to lead Saint Athanasius, Venice was unhappy and instead used the Palaiokapas donation to found San Giovanni (or the Palaiokapas College) in Padua.

The Orthodox Greek Church, which disapproved of Catholic instruction for Orthodox young men, viewed these colleges with suspicion. Already in 1593, the powerful Greek Orthodox community of Venice had founded a school for its children, but it still lacked a higher education establishment that would be on a par with the Catholic colleges. The Greek Thomas Flagginis (1578–1648), a lawyer of the Serene Republic of Venice, offered to finance an Orthodox college in Venice. The Venetian Republic, although tolerant of Orthodox dogma, hesitated twenty years before giving its agreement in 1661. It had already accepted in 1653 the foundation of an Orthodox college in Padua, financed by Ioannis Kottounios (1572–1657). The three Greek colleges under Venetian control—San Giovanni, Flagginis, and Kottounios—would constitute a link between higher education in Italy and the Greek Orthodox world. The last two, especially, became preparatory schools for the University of Padua, educating the future scholars of Orthodoxy.

Despite reservations in some Orthodox circles, wealthy Greek families of the Ottoman Empire did not have an ideological problem with sending their children to study in Padua. As we will see in the next chapter, a future patriarch of

Constantinople, Cyril Lucaris, was a student at Padua, followed by many other future patriarchs and bishops of the Orthodox Church. By means of these studies, Orthodox theologians would be influenced by Italian tendencies and would carry on analogous debates within Orthodoxy. Cases of conversion to Catholicism were very rare, which comforted the Orthodox Church and allowed it to tolerate this kind of study.

Apart from these Greek colleges in Venice, the Greeks participated directly in Italian culture in the Venetian territorial possessions. In 1561 the Cretan Venetian Fransesco Barozzi founded in Rethymnon the Academia dei Vivi, which adopted an Italian tradition: an academy of scholars who met to discuss topics and give lectures, including ones on science. Barozzi, first educated in Crete, had studied philosophy, mathematics, and astronomy at the University of Padua; he published a *Cosmography* in Venice in 1585. But initiatives like Barozzi's remained on the margins of the Greek Orthodox world, where there was no analogous cultural movement. Scholarly associations were never adopted by the Orthodox tradition, and the word *academy* referred only to schools.

The influence of Venetian possessions on the cultural (and thus scientific) progress of Orthodox society was preeminent during the sixteenth and seventeenth centuries, and almost half of the educated Greeks in this period came from these lands. One of the rare schools of the sixteenth century was that of Heraklion (Candia) in Crete. It trained Orthodox scholars such as Meletios Pegas, who became the patriarch of Alexandria, and Maximos Margounios (1549–1602), who became bishop of Cythera. Both of them later taught in Constantinople and in Venice. This era saw the creation in Constantinople of a new Greek aristocracy that would furnish the Sublime Porte, the court of the sultan, with most of its civil servants and technocrats, the men of sciences and of letters that the empire needed to function.

Byzantines and Latins in the Sixteenth Century

The century that followed the fall of Byzantium may be described as "Italian" because most Orthodox scholars fled to the West, usually to Italy, where they were more or less integrated into Renaissance culture. Yet in the sixteenth century a community of Greek scholars linked to the Orthodox world reappeared and tried to formulate a new discourse on the sciences. This community was torn between the militant anti-unionism of the great majority of the Orthodox Church and its privileged links with Italian science.

In the sixteenth century, a humanist spirit traversed educated Greek society

CHAPTER TEN

A Rebel Patriarch

Cyril Lucaris and Orthodox Humanism in Science

The sciences and secular learning in general did not figure among the preoccupations of the Orthodox Church from the fall of the Byzantine Empire in 1453 to the start of the seventeenth century. In fact, for a century and a half, the Patriarchate of Constantinople had a policy of teaching only what was useful for the renewal of the ecclesiastical hierarchy. The only organized Orthodox school in the Ottoman Empire was the Patriarchal School of Constantinople, refounded in 1454; with regard to curriculum, it had nothing in common with its predecessor in the days of the Palaiologos. As we saw in the preceding chapter, other Greek schools, created by Greeks of the diaspora in Italy or in the Venetian possessions, were to be found outside the Ottoman Empire.

The Ottoman Empire based its domination on the organization of millets, a system for controlling non-Muslim populations in which authority was delegated in large part to their religious leaders, appointed by the sultan. The first millet to be created was that of the Orthodox Christian Church, just after the conquest of Constantinople in 1453, when Mehmed the Conqueror gave the privilege to Scholarios. This was followed by the Armenian millet and the Jewish millet. The millets had their own laws (e.g., when a member of a millet committed a crime, the law applicable was that of the millet of the person harmed), but when a Muslim was involved, then *sharia* trumped everything. Millets collected their own taxes in compensation for their loyalty to the empire, and they managed their own educational system, which largely liberated Orthodox communities from Islamic influences.

The leader of the Christian Orthodox millet was the patriarch of Constantinople, who also gained control over the Bulgarian and Serbian churches, which had been independent during the Byzantine period. Thus, the power of the church with respect to education and science increased, compared to what it had possessed in the Byzantine era, when it had to share this responsibility with the

Byzantine emperor. As had been the case since antiquity, the scholarly language of Orthodoxy was Greek, and so education was dispensed in Orthodox schools in Greek. Thus, concerning science specifically, the policy adopted by the Patriarchate of Constantinople affected all Orthodox peoples (whether or not they were native Greek speakers) of the vast Ottoman Empire, which at its apogee extended over parts of three continents: Europe, Asia, and Africa.

Until the start of the seventeenth century, only a few Christian Orthodox thinkers—not more than fifty during the sixteenth century—were involved with the sciences in the Ottoman Empire. Because organized schools did not exist in the empire, we do not find the usual debates on the nature of learning and on scientific teaching. Nevertheless, as we saw in chapter 9, discussion of the validity of secular learning of a humanist tenor had already taken place in the sixteenth century, among Greeks who were in contact with Italian culture. However, the creation of a scientific community worthy of the name among Orthodox Christians of the Ottoman Empire would be a long and slow process; it would not be fully achieved until the century of the Enlightenment. Nevertheless, such a community began to make its presence felt already in the seventeenth century, with a movement that modern historians call "religious humanism."

Cyril Lucaris, the leading humanist in the Orthodox world, was born in 1572 in Crete, which was then under Venetian domination. He studied with a famous tutor in Heraklion, Meletios Vlastos, and continued his studies in Venice and then at the University of Padua. In 1593 he found himself in Alexandria, where his uncle, Meletios Pegas, was patriarch. After taking religious orders, he was sent to Poland in 1594–96 (and then again in 1601–2) to combat the influence of the Catholic Church, which was trying (with success) to convert Orthodox believers to Catholicism. Some years previously, in order to strengthen the Russian church against such a danger, the Constantinople Council of 1593 had founded the Patriarchate of Russia. During his stay in Poland, Lucaris realized the importance of Jesuit colleges, and analogously he organized the Orthodox Academy of Vilnius. At the death of his uncle in 1601, he became patriarch of Alexandria, at the age of twenty-nine. His studies in Italy as well as his trips to Poland inspired him to build up a significant library of European books. It was around this time that he formed significant relationships with the ambassadors from Protestant countries in Constantinople and with theologians of the Anglican Church. In fact, the patriarchs of Jerusalem spent a great part of their time in the capital, near the centers of power: the Sublime Porte, the Patriarchate of Constantinople, and the Phanar (the district of the Greek aristocracy of Constantinople). It was the era when Catholics and Protestants were engaging in a war of influence over gated in the Orthodox world was not that of his Italian professor but a "direct line" Aristotelianism that drew on the tradition in the Greek language, starting with Alexander Aphrodisieus. In fact, Korydaleus was opposed to the theological interpretation of scholastic Aristotelians, insisting on the autonomy of philosophy from theology.

Korydaleus did have to respond to the questions that Christianity posed to Aristotelianism, essentially about the genesis of the world. Here he made a distinction between the Aristotelian genesis, an action precise in space and time, having a reason for being, and genesis as the creation of the world, which is contrary to Aristotle's physics. In short, once the world is created, it is Aristotelian, and philosophical thought is sharply distinct from theology. Aristotle cannot be interpreted in a metaphysical way. Therefore, the scholar's primary duty is to make a distinction between theology and philosophy, keeping the two separate. This position is illustrated by the following anonymous passage, written in 1669 at the latest: "What is true for theology is also true for philosophy, and what is false for philosophy is also false for theology. For theology is based on the light of faith, and philosophy on the light of nature."[3]

Korydaleus wrote two voluminous books on the philosophy of nature: the *Introduction to Aristotle's Physics* and *On Genesis and Decay according to Aristotle*. To these should be added *On Aristotle's Whole Logic*. These books in their printed versions are from 450 to 700 pages each. Despite mentioning only Aristotle in their titles, these works present the whole natural philosophy of the ancient Greeks. After explaining their various theories—from Pythagoras to Simplicius—the books pose questions and then resolve them. The style is rather scholastic, and there are repetitions and rhetorical turns of phrase, but everything is there. Korydaleus's ambition was to offer a panorama of Greek natural philosophy as if the Christian religion had never existed. Thus, in these voluminous tomes Korydaleus did not present the Christian critique of the ancients' conception of the world. He wrote as if philosophy had no theological implications. In the preface to the *Introduction to Aristotle's Physics*, Korydaleus divided philosophy into three parts (referring to Plato as well as the Stoics): physics, ethics, and logic. Physics speaks of the world and theories about it; ethics is concerned with human life, and logic with reason. "As for science about God, it does not enter into the parts of philosophy, but it was called by [the ancients] wisdom par excellence."[4]

Because physics is independent of theology, Korydaleus could allow himself to teach the former without any reference to the latter. He presented reasons for the utility of this physical science, which are both practical (e.g., applications to agriculture) and theoretical. When it comes to the nature of the universe or of light, utility consists of the acquisition of wisdom, for mortals approach wisdom in this way and become men of science, which is pleasing to God.[5]

Speaking of the world, of matter, of genesis and decay without presenting the Christian theory of nature as it had been formulated by the Hexaemerons of the church fathers was something new in the Orthodox world of the Ottoman Empire. Moreover, Korydaleus's tomes were designed for use in the very Orthodox Greek colleges. Korydaleus was not just anybody; he was close to the highest spheres of the church during the reforming patriarchate of Lucaris, and he had taught at the very heart of Orthodoxy, at the Patriarchal School. Given the stakes, the reaction in Orthodox circles to Korydaleus's kind of teaching was rather moderate. By the end of the seventeenth century, Korydaleus would be accepted as a great scholar by the Orthodox world; his name became a reference in the domain of the sciences.

He did, however, have critics who defended the supremacy of theology over philosophy. The leading voice from this camp was Nikolaos Koursoulas (1602–52), originally from Zante (a Venetian possession), who, criticizing the "novelties" of Korydaleus, interpreted Aristotle through the scholasticism of the Middle Ages by arguing that faith is actually the rule of philosophy. In contrast to Korydaleus, who (following Alexander Aphrodisieus) maintained that the goal of the philosopher was to study the whole, Koursoulas followed Simplicius in maintaining that the philosopher should study only "simple" bodies—for example, the sky. Thus, a major part of learning must remain theological. This debate between Aristotelians such as Korydaleus and scholastic Aristotelians became generalized around the middle of the century in the Orthodox world, and as a result, until the end of the century conservative circles were opposed to the Korydaleusian style of teaching natural philosophy. It should be noted that the cause of this opposition was not the "new science" (as in Catholic Italy during the same period) but rather the very teaching of the philosophy of nature. Korydaleus's lack of reference to the Christian values of the Hexaemeron must have exacerbated the conservative antagonism. According to these circles, believers should not get involved in these secular questions as taught by philosophers—and, anyway, they had already been resolved by the church fathers. Moreover, it was unthinkable to teach students theories of nature without discussing at length the creation of the world by God.

Despite these reactions—expected and logical—from Orthodoxy, the official church did ultimately adopt the Aristotelianism of Korydaleus. This rapprochement was facilitated by the thesis of the "double truth" (philosophical truth illiterate and superstitious, the patriarch wrote that they did not need professors such as Copernicus or Galileo because they were inclined to overthrow sacred texts by sophistry.[11] The "debate" ended there.

Until the nineteenth century, then, there was little to distinguish between the positions of church and secular scholars. Debates and discussions traversed Orthodoxy itself: every tendency was present, even among ecclesiastics in the upper hierarchy. In this context, it is not surprising that the first book to present the heliocentric system in detail was written by Michael Mitros, called "the geographer" (1661–1714), future metropolitan of Athens under the name Meletios. While he was a teacher at Jannina (c. 1687–92), Mitros wrote a huge manuscript (of which the oldest conserved copy dates from 1700) in which he described the Copernican system and presented all the new discoveries that Galileo had made with the telescope—sunspots, the moons of Jupiter, comets—and discussed the existence (or not) of heavenly spheres.[12] Nine manuscripts of this text have survived, which suggests that it enjoyed widespread use in Greek colleges. As a man of the church, Mitros could not be a partisan of the Copernican system, but as a cultivated man, he realized that the new phenomena observed by Galileo could not be explained by the geocentric system. Therefore, he supported the geoheliocentric system of the Danish astronomer Tycho Brahe (1546–1601): the Earth remained at the center of the world, while other planets orbited the sun which circled the Earth. This position had long before been adopted by the Jesuit astronomers because it was believed to be in accordance with both the new observations and the Bible.

Thus, what characterizes the reforming current of seventeenth-century Orthodoxy is once again a "return to sources," but this time the sources are not the church fathers and ecclesiastical Byzantines, but emphatically the sources of secular philosophy. After two centuries of uncontested Ottoman domination, the Orthodox communities of the empire were becoming more powerful by controlling essential economic activities. This would give them the vague desire for independence, which would become concrete only after the middle of the eighteenth century, with the advent of nationalism. However, these impulses toward independence assumed ideological expression, such as the affiliation of Orthodoxy with Greek culture in the form of religious humanism. Orthodox believers would thus feel strengthened in the face of both the Ottomans, who had the political power but whose science left much to be desired, and westerners, who were developing a new science that would put them at the forefront of world scientific and technological achievement.

This return to Hellenic sources, which in science was translated into the introduction of natural philosophy and mathematics to Orthodox schools, had a major theological impact: relaunching discussion of the validity of science in relation to theology. But this renewed discussion took place in a new context. Now it was a matter of affirming the glorious past of a people who felt themselves currently "in decadence," that is, subjugated by a non-Christian state. Until then the only reference of the Orthodox had been Byzantium. But now European humanism revalorized ancient Greece and offered a frame of reference that had been neglected after the victory of the anti-unionists. In Aristotelian natural philosophy of the Korydaleus variety, the church fathers were rarely cited and exegetical texts such as the Hexamerons were almost omitted. Korydaleusian science did face two kinds of opposition, from the scholastic Aristotelians and from Orthodox conservatives who had reacted against the introduction of natural philosophy to the orthodox colleges. But most of the later patriarchs would accept Korydaleus's kind of education as beneficial to Orthodoxy because, in the last instance, it was preferable to the new Western ideas of the scientific revolution. In fact, the aggressive Aristotelianism of fervent Orthodox believers such as Koressios seemed to derive from a distrust of novelties coming from both Catholic and Protestant Europe. After all, Aristotle was Greek, and he had been abundantly studied and commented upon by Orthodox Byzantines.

CHAPTER ELEVEN

Toward Russia

The Slavo-Greco-Latin Academy and the Patriarchate of Jerusalem

Since the formation of the Slavic alphabet by Cyril and Methodius in the ninth century and the Christianization of the Rus in the tenth century, the intellectual influence of Byzantium on Russian culture had been primary. One might therefore expect that, along with other cultural and religious aspects, the Russians would be influenced by Byzantine science throughout centuries of relations with the "homeland of Orthodoxy." Although one can detect such influences, these are rather weak, because in Russia science was not seriously cultivated until the seventeenth century. Unlike in Byzantium, the ancient Greek scientific corpus was almost unknown in Russia until then, though books on nature (such as *brontologia* or *selenologia*, describing in a simplistic way meteorological phenomena or the phases of the moon) had significant diffusion after the fourteenth century. Later on, in contrast to what happened among the Orthodox of the Ottoman Empire who went abroad to study beginning in the fifteenth century, the first Russian to obtain a European diploma, at the University of Padua, did not do so until the end of the seventeenth century. Until recently, Russian historians of science accepted the Russian author Alexander Pushkin's apothegm that "the influence of the Mongols, who were Arabs without Aristotle and without algebra, contributed nothing on the level of proto-scientific concepts and interests"; historians of Russian science began their accounts with the eighteenth century.[1] However, this severe judgment was advanced in order to condemn the backward role of the church and to valorize Peter the Great's reforms; it has been moderated by recent research.[2]

In fact, the Russian church, very sensitive to the mystical tendencies of the Orthodox Church, did oppose scientific culture, whether Byzantine or other. Scientific education was rarely offered in Russian monasteries, which acquired great influence under the Tatar régime of the thirteenth and fourteenth centuries. After the fourteenth century, Hesychast teaching had a great influence in Russia, which was propitious terrain for welcoming such a movement. An example of the Orthodox fundamentalism of the Russian church is the banishment of the Latin language. Indeed, for several centuries the Russian church considered Latin (and its documents) to be diabolical. So it is not surprising that the only man of the Renaissance period who was influential in Russia was Maximus the Greek (c. 1480–1556), a prolific translator of ecclesiastical texts into Russian. Nor is it surprising that the cosmology that prevailed in Russia was that of the school of Antioch (see chapter 2) and that the mysticism of a Cosmas Indicopleustes remained in vogue in the seventeenth century.

This withdrawal from (and negation of) scientific learning would gradually be modified in the course of the seventeenth century. Here again, the initiative came from the Russian Orthodox Church. Patriarch Nikon (1605–81) undertook a great battle to modernize Russia by trying to purge the church of any mystical and magical element. At the same time, a debate over the control of the Russian church was taking place, between him and Tsar Alexis (r. 1645–76). It was a "battle of giants" between the two supreme authorities of the vast kingdom. During this period, Russia acquired the Ukraine (1654), where the metropolitan Petro Mohyla (1633–46), in order to counter the Jesuit penetration, had founded the Theological Academy of Kiev, where Latin was taught. In this context favorable to change, Paisios Ligarides arrived in Moscow.

Prophets and Science

Paisios Ligarides was a Greek Catholic from the island of Chios who had studied at the College of Saint Athanasius in Rome under his compatriot Leo Allatius (see chapter 9), who trained him as his successor. Very brilliant, he acquired the reputation of the most cultivated man of the Ottoman Empire. In 1647 he converted to Orthodoxy and in 1652 became bishop (metropolitan) of Gaza, where he stayed only two years. He then went to Bucharest, where he negotiated with officials for permission to go to Russia to participate in Russian ecclesiastical affairs, specifically the struggle between Patriarch Nikon and Tsar Alexis. Meanwhile, the patriarch of Jerusalem, Nektarios, learning that Paisios had not broken off relations with the Vatican or with the Catholic Allatius and that he was not refusing Rome's money, removed him as metropolitan, something that did not prevent Paisios from continuing to bear this title when he addressed the Russians. At Bucharest he seems to have taken part in Nikon's reforms, but on arriving in Russia in 1662, he took Tsar Alexis's side. It remains unclear whether he did so out of personal interest, upon seeing the balance of power swing toward the tsar, or whether he remained at heart a Vatican man who was trying to penetrate Russia by profiting from religious strife among the Orthodox. We do know that he presided over the council (synod) that condemned Nikon in 1666, which gave the tsar a stranglehold over the church of Russia, modeled on Byzantine practice. From a theological standpoint, the council approved Nikon's reforms that imposed Greek worship practices.[3]

Paisios Ligarides inaugurated a policy by Orthodox Greeks toward Russia that would be pursued elsewhere, notably by the Patriarchate of Jerusalem. He saw Orthodox Russia not only as a natural ally of the Greeks but as a force that might help the Greeks liberate themselves from the Ottoman Empire. In 1652 he wrote a book of prophesies (*Chrismologion*), which he dedicated to the tsar, calling on him to liberate his Orthodox brothers. Prophetic literature abounded in the Orthodox world and was generally tolerated by the church. Many priests practiced the art of prophecy, and they were more feared than persecuted. We have seen the predictions of Byzantine astronomer-astrologers about the future of Islam (chapter 3); this time, they were about the future of the Ottoman Empire. Ligarides' predictions for Russia remained in vogue well after his death. Even encyclopedists such as Konstantinos Dapontes (d. 1784), secretary of the princes of Moldavia, believed them as gospel. But, having lived long enough to see the dream of Russians reconquering Constantinople recede, Dapontes turned to the interpretation of the Apocalypse of Saint John and predicted that the Byzantine Empire would never be restored and that the Russians would never reign over Constantinople.

In addition to his prophetic views about the reconquest of Constantinople, Paisios tried to promote Greek culture in Russia. Greek was the language of scriptures and of Orthodox texts, and hence it was indispensable for discussing theological matters. He informed the tsar that the heresies that abounded in Russia derived from the fact that there was no Greek school or library there. "If I were asked what are the supports of the Church and the monarchy, I would reply that in the first place is the Greek language, and second, Greek schools, and third, learning the Greek language," he proclaimed.[4]

As we have seen, the upper clergy often came from wealthy families that had both the means and the desire to pursue studies at the university level. The path to power for the Orthodox within the Ottoman Empire was through the church, which, thanks to the system of millets, exerted an influence well beyond the limits of the spiritual. As in Byzantine times, one could accede to very high posts while continuing to live a life that was quite secular. Thus, certain high clergymen seemed sometimes to take liberties with dogma, which was astonishing in the context of a traditionalist society. Such was the case of Paisios, whose involvement in the occult sciences he did not even try to conceal.

When Paisios was named metropolitan of Gaza, he gave a dozen sermons in Jerusalem before the patriarch Nektarios. Paisios began by comparing the twelve signs of the zodiac with the twelve major festivals of Christianity, showing off his knowledge of astrological signs. In his *History of the Condemnation of Patriarch Nikon*, he gives a description of Nikon that is based on astrology, palmistry, physiognomy, and dream interpretation.[5] When the tsar needed his erudition to furnish scholarly arguments for condemning Nikon, he did not pay attention to the accusations from Jerusalem against Paisios. However, the successor to Nektarios, Dositheos, confirmed the condemnation of Paisios by his predecessor, and this time the condemnation was pronounced not on account of his penchant for the Vatican but for "crimes that modesty does not permit naming."[6] In fact, during the trial against Nikon, the latter had accused Paisios of not following the tradition of the Orthodox fathers and of practicing magic and astrology. Once Paisios had served his purpose, the tsar no longer protected him, and he had to implore the pardon of Dositheos. The patriarch sent him an indulgence in 1670 but withdrew it a year later. Now persona non grata in the Ottoman Empire, Paisios left Moscow for Kiev, where he taught philosophy at the Theological Academy founded by Mohyla, dying in disgrace in 1678. Paisios did not live to see his dream, the founding of a Greek school, realized. But shortly after his death, under Theodore III Alexeyevich's reign (1676–82), such a school was indeed founded, teaching the liberal arts among other things.

From the moment when Russian power decided to follow a policy of expansion in the East, claiming to be the heir of the Byzantine Empire, it needed the learning of the Orthodox Greeks. Their erudition in both theology and secular knowledge was indispensable to accomplish such a project. The Greek church saw in its Russian Orthodox brothers the potential support it desired. The new patriarch of Jerusalem, Dositheos, especially sought Russian support against the Catholics for control of holy sites.

Precisely at this time, a Greek community was developing in Moscow, partly the result of the growing commercial relations between Russia and the Greek communities of the Ottoman Empire. Between Russia and Mount Athos, there was a trade in manuscripts and relics, and the Iberian monastery of Athos acquired the right to found an annex (*metochion*) in Moscow.[7] At the same time, the Greeks played an important role in this city's press.[8] This context proved favorable for the founding of an educational institution at an advanced level, where Greek would be taught and the curriculum would include the liberal arts. The Greek

one also recommended by Patriarch Dositheos: Nicolas Spathar Milescu (1636–1708). Spathar was born in Moldavia of a Greek father. He had studied in Constantinople with Gabriel Vlassios, who was later the bishop of Arta and Lepanto, and then in Italy, probably at the University of Padua. Returning to Moldavia, he mingled in court intrigues, and when things turned out badly, Prince Stephanita condemned him to have his nose cut off.[12] Spathar went into exile in German lands and then in Sweden, where as an expert in Greek and in orthodox theology, he offered his services to the French ambassador Arnaud de Pomponne, supporting him against the Calvinists over the question of the Eucharist. At this time, the Calvinists were using the *Confession of Faith* of Cyril Lucaris (see chapter 10). But because power in Moldavia had shifted in the meantime, Spathar tried his fortune again there but chose the losing party. So he was once again exiled, this time to Moscow, where he was introduced by Patriarch Dositheos to Tsar Alexis Mikhailovich. The patriarch saw Spathar as his man in Moscow, capable of persuading Tsar Alexis to support his policy.

Spathar quickly demonstrated his abilities; he linked himself to powerful men, such as Prince Basil Vasilevich Galitsin and Artemion Sergeyev, and took important posts to serve Russian expansionism into the Balkans. An adventurer in the service of kings, Spathar always remained a man of sciences and letters. In Moscow, he wrote a book on arithmetic, a Greek-Latin-Russian dictionary, and part of a book to educate the young tsarevich, and he translated and adapted into Russian the book of prophecies by Paisios Ligarides. This last book earned him the church's disapproval, but fortunately Dositheos came to his aid. In return, Spathar helped Chrysanthos to conduct his Muscovite business. The latter's mission seemed crowned with at least partial success, with the organization of a group of scholars who translated several Russian works into Greek during his stay: books on the history and geography of Russia and the Far East and the lives of Russian saints. Six years after the foundation of the Slavo-Greco-Latin Academy, Russia at last possessed capable translators.

Chrysanthos also managed to convince the tsar to dismiss the Leichoudis brothers. Following tradition, he had them imprisoned in the Novospaskij Monastery. Spathar himself would have been the man to replace them; a scholar in arts and sciences, he knew Latin and Russian well. Spathar, approaching the age of sixty and not having the same access to the young tsar Peter as he had had with his father, Alexis, no doubt wanted to occupy this post. Nevertheless, Dositheos, as a result of his clash with the Leichoudises, had come to distrust scholars who preferred Latin to Greek.[13] Thus, the Leichoudises were replaced by two of their students, Nicolas Semenov and Theodore (Feodor) Polycarpov, who unfortunately were not up to the task. The brothers' dismissal marked the decline of the academy, since, despite efforts to find good teachers, the Leichoudises remained the most able professors, and so, twenty-two years after his exile (in 1716), the old Sofronios Leichoudis was once again summoned back to the academy.

However, a few years later the Slavo-Greco-Latin Academy would be eclipsed by the Academy of Science of St. Petersburg, founded in 1725, which ultimately did not have a better fate. Even under pressure from the powerful monarch Peter the Great, Russian society continued to resist the teaching of science. Before the creation of the Science Academy, Peter sent dozens of young Russians to study in Europe, but the church and the aristocracy saw only the potential for contamination by seditious Western ideas. It would take another half century for a scientific culture to be truly established in Russia.

In 1675, some twenty years before the arrival of Chrysanthos in Moscow, Spathar, then at the height of his diplomatic career, was sent as the tsar's special ambassador to the young Chinese emperor Kangxi (r. 1662–1725). At the head of an important delegation, Spathar had the triple mission of settling border problems provoked by the adherence to Christianity of Prince Gantimur (of Siberia), of resolving the question of titles and language for the tsar's communication with the emperor, and especially of creating commercial relations between the two countries. On his way to China, he paid a visit to Gantimur, a Siberian prince, to renew the tsar's support, and he surveyed and verified the navigability of the rivers connecting Siberia with China.

Little more than a year after his departure, Spathar arrived in Peking, where he waited only a month to be received by Kangxi. The interpreter for the emperor was the Flemish Jesuit Ferdinand Verbiest, who communicated with Spathar in Latin. Verbiest was at this time the head of the Jesuit mission in China and, at the same time, director of the very prestigious imperial astronomical bureau; he was close to the young emperor, to whom he had rendered various services, using his knowledge of European science and technology as well as his own ingenuity. Verbiest's objective (after restoring the influence of Jesuits in China when the young emperor got rid of his regents in 1669) was to ensure a new route between Europe and China to overcome the difficulties of maritime traffic and the customs barrier of Portuguese-controlled Macao. Spathar's arrival was timely: it potentially enabled Verbiest to obtain the tsar's permission to transit through Siberia. However, Verbiest had everything to fear from a breakthrough by the Russian Orthodox into China, an empire that he was trying to convert to Catholicism. For Spathar, the difficult balance was to knit good relations between Russia and China while still using Verbiest as an intermediary. When Spathar asked him for a Latin-Chinese dictionary, Verbiest lied and pretended that nothing of the sort existed. Moreover, he slandered the Russians to Kangxi, telling him they were heretics and comparing them, vis-à-vis the pope, to rebel Chinese subjects vis-à-vis the emperor.

The Jesuits relied on science and technology to penetrate China. By demonstrating their capacities in this domain, they hoped to make themselves indispensable and to obtain privileges for their religion. Verbiest tried to use this same tool with the tsar, with Spathar's mediation. Instead of sending him a dictionary, he sent a manuscript addressed to Alexis in which he described the exploits of the Jesuits in China in the realms of astronomy and technology. In the introductory letter, Verbiest offered the tsar analogous services in return for his tolerance of the Society of Jesus.[14] For his part, Spathar, a man of science, could not help being impressed by the Astronomical Bureau and its prestigious observatory with imposing instruments copied from those of Tycho Brahe (but finely decorated with Chinese dragons), as well as the scholarly discourse on technology and the sciences contained in the manuscript he was transporting.

But when Spathar returned to Moscow in January 1678, his protector Alexis was dead and his eldest son, Theodore, was on the throne. Spathar hurried to show the new tsar that his mission had been successful, submitting a report in which he described the whole trajectory from Tobolsk, the capital of Siberia, to the Chinese border, as well as his mission in China. As he could not disseminate this report, supposed to be a state secret, he soon wrote a huge book called *Description of Asia*.[15] This book made a great impression on his contemporaries and was one of the books that Chrysanthos paid to have translated into Greek. On the basis of Spathar's books and oral descriptions, Chrysanthos himself wrote a book on China.[16]

Although the Orthodox world showed great interest in Chinese history and geography, it paid little attention to Verbiest's science and technology. Yet the manuscript brought back by Spathar was impressive enough. Without mentioning cosmological questions, it described the organization, tasks, and instruments of the Chinese Astronomical Bureau, giving details on scientific instruments, mechanical clocks, and magnetism, along with methods of leveling and of transport with the help of pulleys. It also demonstrated Verbiest's skill in the manufacture of canons. The science and technology presented in this manuscript were nothing new for a western European scholar; they were even outmoded as concerns astronomy and its instruments, given the creation of the observatories of Paris and Greenwich (respectively in 1667 and 1675). Nevertheless, nothing like it existed in the Orient—especially scientific instruments.

Why this lack of interest? It is unlikely that Spathar (desiring to please the new regime) did not actually give the manuscript from Verbiest to the tsar and his entourage. And it would have been foolish to keep to himself a letter addressed to the sovereign, even a deceased one, for the danger of discovery would be great. The answer lies in the Russian society of this period. Apart from the Leichoudises, Spathar was the only scholar who knew science at a level sufficient to understand Verbiest's text. It would not be until Peter the Great's "Grand Embassy" in 1697 that the country turned (forcefully) to the science and technology of western Europe. But another problem was insurmountable. Verbiest was a Jesuit and did not hide his true purpose, which was to win toleration for the Society of Jesus in Russia. This is what he had written (in Latin) to Alexis:

> And because Your Majesty the Tsar has appointed me as his Latin-Chinese interpreter, I will offer now to Your Majesty the Tsar, as "books of request," a Latin translation of these Chinese volumes which I have recently offered to the Chinese-Manchu Emperor, in which I sketched, by means of figures taken from Mechanics, the restitution of [Western] astronomy in China by myself. . . . But [my intention is] to abundantly fulfill my duties as an interpreter, asking [in return], together with my fellow fathers who are residing here [in Peking] with me and are very dedicated to the obedience of Your Majesty the Tsar, for only one thing, viz., that You will always look at our Mother, the Society of Jesus, which since our infancy taught us all these languages, with the same eyes, as Piety itself and our Society are praying for Your Majesty the Tsar's happiness and eternal years of the [Russian] Empire.[17]

Only a person unfamiliar with Orthodoxy could have penned such a letter. The Orthodox Church viewed the Jesuit missions with enormous distrust, and the Jesuit fathers who traveled to Russia had to live in assigned residences. Since its defeat in Poland, the Orthodox Church had been on the defensive; moreover, it much preferred compromising with the Protestants than with Catholics—and, on top of that, the Jesuits were still close to the Vatican. The Russian church saw Jesuit science as a Trojan horse being used to penetrate the kingdom.

Chrysanthos Notaras did not have the same problems. A cultivated man with a curious mind, he was interested in all the sciences, wherever they came from. When Spathar (probably shortly after Chrysanthos's arrival in Moscow) showed him the manuscript (either a copy he had made for himself or the original located in the library of the tsar, to which he had access), he was astounded. The Chinese xylographies of monumental instruments of astronomy, the very lively account by Verbiest of his combat against the Muslim astronomers to prove the superiority of European astronomy, his exploits in mechanics, the drawings of measuring instruments and surveying methods—everything fascinated Chrysanthos. In fact, it was his first contact with the world of the new European science, which until then had been despised by the Orthodox world. And, again, these were not the most recent discoveries or ideas. In cosmology, Verbiest, like his Jesuit brothers, remained a Tychonian, for this system could explain the phases of Venus (as it turns around the sun) as well the existence of satellites of Jupiter (as all heavenly bodies do not circulate around the Earth). On instruments, Verbiest was behind by several decades in comparison to Europe, for he constructed huge instruments, hence subject to distortion, and did not employ the eyepiece as an aiming device. Nevertheless, what he did show seemed a marvel to Chrysanthos. The Ottoman Empire did not possess an observatory, and the only attempt at one, made by Muslim astronomers in 1577, had not lasted very long, because of religious reactions.[18] Byzantine scientific tradition did not include an instrumental culture, either, and the only texts concerning instruments were on the theory behind the astrolabe, so the only scientific instruments that Chrysanthos had previously seen or read descriptions of were small astrolabes. It is not surprising that he immediately commissioned a copy of Verbiest's Latin manuscript. It was undertaken by a student at the Slavo-Greco-Latin Academy and finished in January 1693.

Chrysanthos kept his copy in the library of the annex (*metochion*) of the Patriarchate of Jerusalem in Constantinople. In effect, this was an extremely indirect breakthrough by the new science into the Orthodox East, because it was accessible only to a narrow circle of readers at the library who knew Latin. However, this library was being enriched by a great number of manuscripts dealing with science, thanks to the special interest and collecting spirit of Chrysanthos, who would become patriarch of Jerusalem at the death of his uncle in 1707.

After the end of the seventeenth century, the Patriarchate of Jerusalem—as well as the rest of the Greek world—continued to maintain close relations with Russia, which long dreamed of reconquering Constantinople. But the tsars, having acquired control of the church in Russia, now depended much less on the help of the Greek patriarchates. As for the sciences, after the reforms of Peter the Great the influence of the Eastern Orthodox Church on science education in Russia would continue to decline. Science would henceforth be taught either by Europeans or by Russians who had studied at European universities. In the eighteenth century, the spread of science in Russia would become an affair of state.[19]

CHAPTER TWELVE

Who Were the Heirs of the Hellenes?

Science and the Greek Enlightenment

In 1697, some years after his mission to Moscow, Chrysanthos Notaras, a monk, was sent by his uncle Dositheos, patriarch of Jerusalem, to Padua in order to complete his education. The curriculum of the University of Padua included theology courses that were attended not only by Catholics but also by Orthodox students such as Chrysanthos. This might appear to contradict the hatred of Catholics felt by a majority of the Orthodox clergy, but the clerical aristocracy needed men who well understood the dogmas of the "Latins." Following a Byzantine tradition, a certain portion of the clerical aristocracy maintained good relations with the Catholic Church. Chrysanthos's professor at Padua was the Greek Catholic Nicolas Komninos. His attachment to Catholicism did not prevent him from considering Italy as a sinful land, and he maintained friendly relations with Dositheos, to whom he wrote that his nephew was leaving Padua more knowledgeable—and without having been perverted by Italian morals.[1] The faculty of arts in Padua was Aristotelian, but Chrysanthos was not content with Aristotelian study; he had a curious mind and knew that a new science had been developing for a long time, if only from the manuscript of Verbiest that he had had copied (chapter 11).[2] In 1700, at the end of his study in Padua, he went to France, "the land of the Celts," according to Komninos.

In Paris, Chrysanthos made the acquaintance of liberal theologians such as Louis Ellies Du Pin, Alexandre Noël, and Michel Le Quien. Still more important for him was his contact with men of science. As he was the nephew of the patriarch of Jerusalem and had traveled throughout eastern Europe, Chrysanthos was well received by French scientists. After asking to visit the Paris Observatory, he was welcomed and lodged there for a week by John Dominique Cassini, first director of this modern establishment. As he himself wrote, "There we observed with him [Cassini] with the aid of the largest telescopes the Moon, Jupiter and its so-called satellite-stars, the galaxy, and other things. He then told us that by

In a report on the organization of the Academy of Bucharest, he called for three professors, the first to teach Aristotle (*Logic, Rhetoric, On Heaven, On the Soul, Metaphysics,* etc.); the second to teach Isocrates, Sophocles, Euripides, Gregory the Theologian, Synesios, Pindar, Xenophon, Plutarch, Thucydides, Demosthenes, and the Letters of the Apostle Paul; the third to teach Chrysoloras, Cato, Pythagoras, Aesop, and Homer. This was the Byzantine humanist curriculum; the time was not yet ripe, at the start of the eighteenth century, to revolutionize science education. First, education needed to be reestablished fully in the Orthodox world.

Separate Philosophy from Theology? Antrakites between Descartes and Euclid

One of the scholars who reestablished science education was Methodios Anthrakites (c. 1660–c. 1736). Anthrakites, after studying at the School of Jannina, took his monastic vows and around 1697 went to Venice as priest at the Orthodox Church of Saint George. During the decade of his stay in Italy, he studied science, probably in Padua. Around this time he must have met and become friends with Chrysanthos. Returning about 1708 to Ottoman Greece, he taught at the School of Kastoria and the "Great School" of Jannina. He wrote an enormous textbook in three volumes, the *Mathematics Course*, which would shape science education in the Greek Orthodox world during the first half of the eighteenth century.[10] This book provides a complete course, detailed and rigorous, of the "mathematical sciences" as they were taught in Padua at the start of the eighteenth century, plus some Byzantine texts. In addition to Euclidean geometry, Anthrakites presented books by Ypsicles and Anthemios, the *Sphere* of Theodosius, geometric constructions and trigonometric tables, logarithms, the sphere of Proclus, two Western-influenced treatises on the astrolabe, instructions for using astronomical instruments such as the quadrant (but not the telescope), theoretical geometry, and pre-Newtonian optics. In astronomy, the Copernican and Tychonian systems were presented—before being rejected by arguments along the lines used by Cremonini (who was professor in Padua during Anthrakites' stay in Venice), while Kepler was not even mentioned.

There was nothing revolutionary in the teaching of this monk and friend of Chrysanthos's. However, Anthrakites seems also to have included—or at least presented—the philosophical ideas of the French philosophers Nicolas Malebranche and René Descartes. No doubt this novelty, and the fact that he clearly stressed the teaching of science and not philology, led Orthodox fundamentalists to accuse him before the Holy Synod of being a heretic. In 1723 Anthrakites was summoned to Constantinople to refute these accusations. Condemned, he was excommunicated and his educational books banned. After Anthrakites confessed his Orthodox faith and ceremoniously burned some of his own manuscripts, the church lifted his excommunication and authorized him to teach again, on condition that he use the course of Korydaleus (see chapter 10). In a novel defense of himself, Anthrakites claimed that the church was condemning him for his philosophical ideas and not for having departed from Orthodox dogma. Thus, he reopened the debate within the church on the separation between philosophy and theology. His stance succeeded in heightening the anger of the Holy Synod, which reaffirmed its position that only the Peripatetic philosophy of Aristotle should be taught.[11]

This controversy bore on general principles and not on the subjects actually taught in natural philosophy. In fact, Anthrakites always remained an Aristotelian, and he perpetuated into the eighteenth century the Orthodox humanist ideas of the previous century, which featured the renaissance of Greek science. His student Balanos Vassilopoulos, who in Venice had edited his teacher's manuscript *The Mathematical Course*, wrote in 1755 a treatise that he sent to the Academy of Science in St. Petersburg in which he claimed—wrongly—that he had found the solution to one of the three mathematical problems of antiquity, that of doubling the cube with the aid of a ruler and compass (Delos's problem).

Enlightened Clerics and the New Science:
Voulgaris and Theotokis

Because of the links between Orthodox scholars and the Greek community of Venice, the introduction of the new science into the Orthodox world followed on the heels of its introduction into the University of Padua, the leading university in the Venetian Republic. Until 1678, the chairs of physics at this university were called *ad lecturam meteororum Aristotelis* (lessons on Aristotle's Meteors) and *ad lecturam meteororum et parvorum naturalium Aristotelis* (lessons on Aristotle's minor publications concerning nature). Evidence of the new physics did not appear until 1715, and it was not until 1739 that the new physics was fully taught by Giovanni Poleni (1683–1761), the chair of *ad mathesim et ad philosophiam experimentalem* (mathematics and experimental philosophy), who created his famous Teatro di Filosofia Sperimentale (theater of experimental philosophy), equipped with four hundred instruments to teach experimental physics. Poleni was succeeded by Alberto Colombo, who was in turn succeeded in 1777 by the

Cretan Greek Simon Strattico (1733–1824), who had studied at Kottounios's Greek College in Padua and would be fired by the Austrians for political reasons in 1798, when the latter took control of Venice. This new chair in experimental physics spelled the end of Aristotelianism in Padua, a change that had repercussions in the Orthodox world.[12]

The first reactions by Orthodox students to these reforms were negative. If they went to Italy, it was usually to study Greek sciences, which were being taught at a rather elementary level in the Greek schools of the Ottoman Empire. But gradually a new discourse arose that aimed to reconcile the existence of the new science with a presumed renaissance of ancient Greek science. This new discourse appeared in the prologues of Greek science books around the middle of the century, some fifteen years after the establishment of the chair of experimental physics in Padua. Until that time, Greek scholars conceived of the history of science as solely about Greek and early Byzantine science. Henceforth, the new European science would be integrated into this history as the brilliant heir of ancient Greek science.

Eugenios Voulgaris (1716–1806), the most influential Greek scholar and cleric in southeastern Europe in the second half of the eighteenth century, was born in Corfu, a Venetian dominion. He studied at Padua, where he followed the first courses in experimental physics given by Poleni, and in 1742 became director of the Maroutsis School in Jannina. There Voulgaris taught Gottfried Leibniz, John Locke, and Voltaire, which exasperated Anthracites' student, Balanos Vassilopoulos, director of another school in the city, who forced Voulgaris to go teach "these insanities" elsewhere. Voulgaris went to Kozane, where he was better paid. When Kozanians showed appreciation for his innovative teaching, he was invited back to Jannina with an even higher salary. In 1753 the enlightened patriarch Cyril V created a school (called the Academy) on Mount Athos and appointed him to direct it. At the heart of mystical Orthodoxy, Voulgaris taught, according the scholarly terminology, the "new science" and the "new philosophy," meaning the ideas of the European "scientific revolution" developed after the sixteenth century. His teaching attracted many students as well as many troubles, which obliged Voulgaris to quit the Academy in 1758. Meanwhile, he had translated and adapted a number of science textbooks that would later be printed, including *Elementa geometriae planae et solidae et selecta ex Archimeda theoremata* by the Jesuit Father Andrea Tacquet (1612–60), based on the 1710 edition by William Whiston.[13] In 1759, Seraphim II, a patriarch who supported the Greek revolt against the Turks, called Voulgaris to head the Patriarchal School in the capital of Orthodoxy, Constantinople. The experiment lasted two years, until Seraphim's fall. The new patriarch, Ioannikios III, and the majority of bishops were hostile to this intrusion of Western civilization into Orthodoxy. Voulgaris would leave again, and after a stay in Romania (then administered by Greek princes named by the sultan), he arrived in Leipzig with the goal of publishing teaching manuals such as his *Elements of Mathematics*, based on a book by Andreas Segner.[14] There he met the Russian marshal Theodore Orlov, who introduced him to Catherine the Great. Voulgaris, disappointed at resistance to new ideas and the intrigues in the circles around the Patriarchate of Constantinople, finally settled in Russia, where he was named archbishop of Slavonia and Chersonesos. In 1779 he left the archbishopric and in 1802 retired to the monastery of Alexander Nevsky, where he died at the age of ninety. In 1776 he was named a member emeritus of the Science Academy of St. Petersburg.[15]

Voulgaris is a good example of the relations between cosmopolitan Orthodox clerics and the Enlightenment. On the one hand, he considered himself as heir of the ancient Greeks and was proud of it; on the other hand, he thought that the Greece of his day—confused with the millet of the Ottoman Empire—was totally decadent and would owe its salvation to the teaching of the new European science. For him, the new science relied on developing the science of the ancients. He considered Diophantus, for example, as "the sovereign of all arithmetical thinking." Nevertheless, "his marvelous invention, the art called algebra, was developed and perfected by François Viète, René Descartes, and others."[16] Similar sentiments were expressed by other Greek partisans of the Enlightenment, who considered the great European savants to be the children of the ancient Greek philosophers.

Voulgaris's adherence to the new sciences, which were accompanied by the philosophical and political ideas of the Enlightenment, was strongly shaken by the French Revolution. A portion of the clergy as well as conservative circles of the Phanar (Constantinople's Greek aristocratic quarter) close to the Sublime Porte feared the impact of republican ideas on Greek supporters of the Enlightenment, not to mention atheistic fallout from the revolution itself. As in Russia, even enlightened people who had favored new ideas made retractions. In 1805 the almost ninety-year-old Voulgaris, then in Russia, gave his assent to the publication of his old manuscripts, including a translation he had done almost fifty years earlier of the fourth part (titled "De systemate Universi") of a book by Fortunatus a Brixia (1701–54).[17] This book presented cosmological systems from Plato to Newton, via the church fathers, Kepler, Gassendi, Descartes, and others, and it came out in favor of the Tychonian system, which retained phenomena while remaining faithful to sacred texts.

Voulgaris's fame as well as his title of archbishop gave great prestige to the science manuals he wrote. In natural philosophy, two of them had some success: *Those Who Please Philosophers*, a compilation from books he had read during his study in Italy at the end of the 1730s that included numerous references to Descartes, Newton, and Leibniz; and a translation he had completed in the 1750s of a book by Johan Friedrich Wucherer (1682–1737), *Institutiones philosophiae naturalis eclecticae* (Jena, 1725).[18]

Voulgaris's books were soon eclipsed by those of his junior Nikephoros Theotokis (1731–1800), who followed a similar career. Also born in Corfu, he, too, had studied experimental physics with Poleni and then mathematical sciences with Eustachio Zannoti, director of the Bologna Observatory. After his studies in Italy, he returned to Corfu, took his monastic vows, and created a school where the new science was taught. In the 1760s he was called by the patriarch to Constantinople, where he was appointed preacher of the Grand Church. His teaching of the new science and philosophy shocked conservative circles of the church and Phanar; so Theotokis went to Jassy in Romania, where the ambience seemed more liberal than in the capital of the Ottoman Empire. He soon left for Leipzig, where (like Voulgaris) he went to publish his manual *Elements of Physics*.[19] In 1773 he returned to Romania, where he directed the Academy of Jassy. Facing resistance to his teaching and considered a revolutionary by both clerical and secular conservatives, he accepted in 1776 an invitation from Voulgaris to join him in Russia, where three years later he succeeded him as archbishop of Slavonia and Chersonesos. In 1782 he became archbishop of Astrakhan and Stavropol. In Russia, he engaged in polemics against Old Believers, those who had refused to follow the reforms of Patriarch Nikon in 1652, and he worked to end this century-old discord by promulgating the principle of Edinoverie (unity of faith). In 1792 he retired to the monastery of Saint Daniel in Moscow, where he died eight years later.[20]

Theotokis's *Elements of Physics* became the book of reference for the diffusion of new science in the Orthodox world for at least two reasons. First, this textbook was printed as early as 1776 (other books, such as those of Voulgaris, had circulated as manuscripts for several decades before being printed). Second, Theotokis was not content with presenting "novelties"; he taught physics in a rigorous manner that required solid mathematical skills, available in another of his manuals, the *Elements of Mathematics Compiled from the Ancients and the Moderns*. This book was printed in Moscow in 1798–99, but it circulated in manuscript form as early as 1764. The two books together spread mathematical physics in the Orthodox world at a level approaching that of the University of Padua. *Elements of Physics*, like almost all science books written by Greek savants in the eighteenth century, was a compilation of European manuals with some additions. Two principal sources for Theotokis, both no doubt acquired in Italy during his student period, were the physics of Pieter van Musschenbroek (Italian edition, *Elementa Physicae conscripta in usus academicos a Petro van Musschenbroek)*, and the Italian translation of the physics of Abbé Nollet, of which the five first volumes were published between 1746 and 1766. Peter van Musschenbroek's physics was very popular among Greek savants who had studied in Padua with Poleni, a correspondent of the Dutch scholar's. Abbé Nollet also corresponded with Poleni. In order to write his physics manual, Theotokis used Musschenbroek's book as well as Nollet's to complement it. He organized his manual into ten thematic units: general properties of matter, kinetics, mechanics, liquids, optics, heat, aerostatics, acoustics, electricity, and magnetism. Thanks to this manual and the one on mathematics, Theotokis was the first to present differential calculus to Greek schools in a rigorously didactic way. The *Elements of Mathematics* devoted eighty pages to it, using Leibniz's terminology.

New Science and Traditionalist Society: The Case of Moisiodax

A prominent teacher of new science was the Romanian Iosipos (born John) Moisiodax (c. 1725–1800), from Cernavoda, a village on the banks of the Danube. After having followed in Thessalonica the Aristotelian course taught by Iannakos, a sworn enemy of any attempt at modern education and of new science teaching, he sought to enlarge his knowledge by going to Smyrna in 1753, to the brand-new Evangelical School where the curriculum was also totally traditional. For a young man who had already crossed the Balkans at a time when the modernist spirit of the Enlightenment was spreading in the Orthodox world, this was as boring as could be. As another student described the school, "The professor [Ierotheos Dendrinos] and the school resembled all the other professors and schools of Greece, that is to say, they dispensed an impoverished education, accompanied by plenty of drubbings."[21] And so Moisiodax tried to find financial support to pursue study in Padua.

Although this university was the place to study for many bishops and patriarchs of the Eastern Church, Orthodox clergy did not unanimously approve of it. The thousand-year-old debate between pro-Latins and fervent Orthodox believers could resurface at any moment in the Orthodox world. When the young

Moisiodax requested financial aid to study in Padua, the professor of the Evangelical School, Ierotheos, angrily replied: "All those who study in the French [i.e., Catholic] countries become atheists and upon their return they lead others astray."[22]

Surprisingly, the Serbian monk Dositheos Obradović (1742–1811), who would become a crucial figure in the introduction of new educational ideas in Serbia, described Ierotheos as sympathetic to the Enlightenment, praising him as very erudite, the "new Socrates of Greece," and an enemy of the monks who profit from superstitions.[23] Obradović himself remained an Orthodox who provided his Serbian compatriots with exemplary Greek teachers, in order to inspire them to develop a national educational system.[24]

Moisiodax's trajectory in quest of scientific learning and his experience as a teacher illustrate the complex relations between sciences and Orthodoxy at the end of the eighteenth century. Disappointed in the traditionalist education in Smyrna, he went to Mount Athos, probably in 1754, to follow for two years the modern scientific education of Eugenios Voulgaris at the school that had just been founded by the patriarch of Constantinople, Cyril V. At this time, Voulgaris was teaching algebra according to Christian Wolff, the geometry of Andrea Tacquet, and the physics of Musschenbroek, an education quite compatible with (and comparable to) that of secondary schools of several countries in western Europe. Alongside the purely scientific courses and in his effort to free Orthodox education from sterile Aristotelianism, Voulgaris also taught the *Essay on Human Understanding* by John Locke, one of the principal sources of the empiricism that so influenced Enlightenment philosophers. The education offered at the Sacred Mountain of Orthodoxy, which reflected the new science and its methodology, contrasted strikingly with the outdated and unimaginative education offered in the lively merchant town of Smyrna. The old and the new frequently intermingled during the eighteenth century. Among the pupils of Voulgaris, we find future scholars at both ends of the spectrum: Athanasius Parios, an Orthodox fundamentalist who wrote pamphlets full of hatred, and Christodoulos Pamplekis, an Enlightenment militant who broke all ties with the church, refusing to give it any authority whatsoever.

Despite the modern character of Voulgaris's teaching, Moisiodax was not yet satisfied. He thought that the proportion of scientific education should be greater: "Our results in philosophy would be much better if the savant [Voulgaris], weighing exactly the poverty of the situation and the brevity of his associates' time, and also the urgent necessity of the Hellades, would dispense and profess lectures (mainly oral), especially in mathematics and physics."[25] Moisiodax had not abandoned his old dream of going to Italy. A good way for someone who did not have the means was to enter the ranks of the unmarried clergy, which offered remuneration and mobility.[26] Thus John Moisiodax, under his new name Iosipos (Joseph), arrived in the Greek Orthodox community of Venice around 1759 in order to study the natural sciences at the Università degli Artisti in Padua. There he followed the last courses of the elderly professor Giovanni Poleni, who taught experimental physics in his laboratory Teatro di Filosofia Sperimentale, and of his successor Giovanni Alberto Colombo. The instruments of the Teatro introduced Moisiodax to the world of experiments and technology, which was still unavailable in the Ottoman Empire. This world and its methodology were what Moisiodax wanted to incorporate in Orthodox education back home; the circle of his learning was closing, and the circle of his teaching was opening.

Thanks to his comrade at the university, Constantine Karaioannis, who was the personal physician to the prince of Moldavia, Gregory Ghikas, Moisiodax became professor and then director at the Academy of Jassy. In this era, the princes of Danubian principalities, Gregory Ghikas and Alexander Ypsilanti, Enlightenment men influenced by the Encyclopédistes (i.e., the French writers who compiled the *Encyclopédie* edited between 1751 and 1772 by Denis Diderot and Jean le Rond d'Alembert) were trying to introduce the new culture and especially French culture into their lands. Thanks to this ambience, Moisiodax had the opportunity to teach science as he had studied it in Padua. On his arrival at the Academy in the autumn of 1765, he gave a public lecture that was a manifesto in defense of the new science. In Moisiodax's account, mathematics was the spearhead of the new philosophy, which owed its grandeur, on the one hand, to the learning of great scholars—he cited Plato, Aristotle, Leibniz, and Newton—who, since antiquity, had contributed to its development and, on the other, to the idea that advances were no longer based on the irrefutable authority of any of them. This manifesto, calling for a veritable cultural revolution, shocked traditionalist Jassy society, which was composed of local seigneurs, Phanariot princes, the emergent petite bourgeoisie, and Orthodox clergy, none of whom were accustomed to any discourse questioning the authority of the ancients.

Moisiodax's problems were not long in coming. Shortly after the speech, a discussion between Moisiodax and a scholarly member of the clergy turned into a debate on the physics of Aristotle. Moisiodax was immediately accused of Latinophilia and was obliged to defend himself in a statement (distributed in writing to the court, to physicians, and to the nobility of the city) in which he answered his interlocutor's accusation. He especially refuted Aristotle's theory of matter and defended an atomistic theory that held that God moves atoms. Moisiodax's statement went much farther than a debate over science; he attacked the control of the Aristotelians over the church for political and personal purposes and concluded by saying he was not ashamed to say that it was Greece that needed Europe, for it lacked everything while Europe possessed everything.[27]

Moisiodax did not last long at the Academy, preferring to retire (around 1767) to Bucharest, where he earned his living by teaching private courses and writing science textbooks. Nine years later, the Academy of Jassy found itself again without a director, because Nikephoros Theotokis, encountering the same resistance as Moisiodax, was also obliged to retire. Despite these withdrawals by "renovators," Prince Gregory Ghikas continued his efforts to promote change, putting fresh pressure on Moisiodax to return to lead the establishment. So Moisiodax came back to Jassy, where he introduced in his classes a much more recent textbook than Tacquet's, the *Elementary Lessons in Mathematics* (1741) by Nicolas Louis de Lacaille, which had already been successfully marketed in Europe. Despite assurances given by his friends at the court, conservative circles did not approve of an education so oriented to science, and they put such pressure on him that he was again obliged to retire, less than a year after his return. His detractors attacked him on two fronts, a well-known tactic employed by Orthodox fundamentalists: they accused him of being a Latinophile and of teaching lessons for grocers, meaning mathematics. In effect, these fervent conservatives willingly confused mathematics with practical arithmetic.[28]

Although the detractors of Moisiodax employed the religious argument of Latinophilia, their polemic was not just a reaction by Orthodox fundamentalists to any kind of innovation (specifically, the introduction of the new European science). The reaction against Moisiodax came from a whole society that felt threatened by a new culture that represented the irresistible rise of European power, which would soon sweep away the outdated and exhausted Ottoman Empire. At the start of the second half of the eighteenth century, a new wind coming from the West, a tardy current of the European Enlightenment, blew through the Orthodox world. Its disturbing effects forced the Orthodox believers to find a balance between their convictions and their submission to the Muslim power, on which they depended for their privileges. Obviously, the prime condition for keeping those privileges was to prevent any part of the millet from rising up against the Sublime Porte. But the new scientific ideas that criticized the past with an investigative and open spirit, and which demonstrated the scientific and technical superiority of Europe, undermined any spirit of submission. Thus, attempts by some educated princes of Danubian hegemonies or by a few innovative patriarchs in Constantinople to introduce the new science into education met with strong resistance.

Retiring to Vienna, a city with a flourishing Greek merchant community, Moisiodax wrote and published in 1780 a counterattack on this resistance to the new science and, just a year later, a book on geometry and cosmology.[29] It is remarkable that at the end of the eighteenth century, he felt forced to write ten pages in defense of the heliocentric system, which had been the spearhead of the new science since the start of the seventeenth century. Moisiodax explained to the reader that philosophers should advance the most probable ideas, without ever holding them as certain. His position was not the one held by some defenders of the heliocentric system who explained that it was only a convenient mathematical solution that did not necessarily represent physical reality. Moisiodax's caution derived from his revolt against any authority and his thesis that the modern scholar should always doubt his findings and participate in a constant process of research. However, his prudence and his adherence to Orthodoxy are both visible in his discussions of the calendar problem. Despite the evident discrepancy between the Julian calendar and the seasons, the Holy Synod of the Orthodox Church was still not disposed to adopt the "papal" Gregorian calendar, for the Synod would be accused of submitting to that heretic, the pope. Moisiodax defended the Eastern Church to Western detractors who accused it of obscurantism, explaining that the Gregorian calendar had been rejected not out of ignorance but out of caution, because uneducated people were not yet ready to accept such a change, which they felt would mean abandoning Christian faith and traditions.[30]

As in the cases of Voulgaris and Theotokis, Moisiodax was not actually departing from Orthodox dogma. The dissidents who would spread the new European science did not adhere to either Catholicism or Protestantism; they would fight inside Orthodoxy to change mentalities and to adopt modern attitudes to science and technology. This battle would be conducted within the Phanariot milieu of the Danube principalities. Despite his setbacks, Moisiodax returned one last time to the court of the princes of Walachia as tutor to the sons of Prince Alexander Ypsilanti (*hegemon* from 1774 to 1782). The presence of Moisiodax influenced Enlightenment activists such as Rigas Feraios, who popularized the new science, and Panagiotis Kodrikas, who translated and published in 1794 a provocative French book then almost a century old, Fontenelle's *Conversations on the Plurality of Worlds*.[31] Moisiodax's successor at the Academy, Procopios from Peloponnese (who had also studied in Europe), also defended the new science.

After the French Revolution: Popularization of the New Science

The spread of the new science in the Orthodox world did encounter some resistance on the part of the church, but overall opposition was moderate, coming from conservative circles of the ecclesiastical hierarchy as well as from a society that was afraid of change, especially when it derived from the Latins. As we have seen, Voulgaris, Theotokis, and Moisiodax were all forced by such pressure to resign from the schools where they taught. The first two preferred to continue their ecclesiastical careers in Russia; their fame in the Orthodox world was such (particularly after they became archbishops) that the church did not dare to condemn their scientific ideas openly. Thereafter, professors could use the textbooks of Theotokis to teach the new European science, and soon many other books of this kind were published. This change in scientific education was fostered by the fact that after wars between Russian and Turkey (1768–74) had brought about agreements favoring the Orthodox communities of the Ottoman Empire, the merchant classes of these communities had begun developing and establishing themselves in various counties of Europe. The result was a generation of educated Greeks who were directly influenced by the French Enlightenment. Though much less scientifically talented than the generation of Voulgaris and Theotokis, these men worked to propagate European science to the greatest possible number of Christian subjects of the Ottoman Empire. Their goal was now much wider: national emancipation, not just the development of science as such. Most of the new publications were books of scientific popularization based on French or German encyclopedias—often taken from the very *Encyclopédie* of Diderot and d'Alembert and written by educated people who had not necessarily studied science. For those people, the *Encycoplédie* was a main source of inspiration to propagate the European Enlightenment's ideas to the Orthodox world.[32]

The case of Rigas Feraios (1757–98) is characteristic of this movement. Rigas was not a scholar but a politician, who, inspired by the French Revolution, conceived of a Greek-speaking Balkan Republic. One of his first books was an *Anthology of Physics*, which he published in Vienna in 1790, at the same time (significantly) as his translation of a French romantic novel. Although different in essence, both books were of the same order: addressed to an Orthodox audience to whom Rigas offered various aspects of the French Enlightenment spirit. As its title indicates, the *Anthology of Physics* was an anthology of articles on the natural sciences mostly based on the *Encyclopédie*: why it rains or snows, the nature of the galaxy, how an electrostatic machine works, where the heat of the sun comes from (the fashionable answer was "from electricity"), how many satellites Saturn has. The laws of nature explained everything; Rigas included no appeals to the supernatural. The book's conversational language was extremely simple. "My goal," Rigas explained in his prologue, "is to benefit our nation and not accumulate words in order to show off knowledge; I intend to explain with as much clarity as possible so that everybody can understand and acquire a small idea of incomprehensible physics."

The French Revolution and its consequences in the Balkans (the advance of Napoleon, movements of insurrection) led the Orthodox to react more violently than in previous decades to the introduction of new scientific ideas—a reaction that soon became a reaction against any kind of scientific education at all. Shortly after the publication of Panagiotis Kodrikas's translation of Fontenelle's *Conversations*, Sergios Makraios (c. 1740–1819), one of the directors of the Patriarchal School of Constantinople, published an astonishingly backward-looking book titled *Trophy of the Panoply of the Hellades against the Partisans of Copernicus*. In the classic style of questions and answers in an archaic language, Makraios demolished the heliocentric system—not by denying the theory of universal gravitation but by interpreting it in the manner of Aristotle (any matter separated from its natural milieu has a tendency to return to it). Makraios vehemently opposed "Westerners": "The lightweight Fontenelle was foolishly mistaken to think he could reach Olympus by getting mixed up in celestial things. Seeking the plurality of worlds, the crazy Descartes got lost among his whirlpools, maneuvering as he wished."[33] But this tardy reaction could not contain the growing educational movement in favor of the new science. European science in all its aspects was now too well anchored in Greek education.

Take the example of astronomy. In 1803, seven years after Makraios's book was published, a Greek translation of Joseph-Jérôme de Lalande's *Treatise of Astronomy* appeared, enriched with new discoveries such as the small planets found by the Italian priest-astronomer Giuseppe Piazzi on January 1, 1801. Lalande's book, originally published in 1764 (and reissued in 1771 and 1792), was until the start of the nineteenth century a "standard" textbook in Europe, rich in information on astronomical instruments and methods of calculation. The Greek edition was the product of two partisans of the Enlightenment who came from Milies in Thessaly, Daniel Philippides and Anthimos Gazis.

sary than mathematical or scientific classes . . . for what is the advantage for the students who follow these courses to learn figures and algebra, cubes and cubocubes, and triangles and trigonosquares, and logarithms, and symbolic calculations, and projected ellipses, and atoms and voids and whirlpools, and forces and attractions and weight, and qualities of light, and polar auroras, and optics, and acoustics, and thousands of similar and monstrous things, in order to count the sand on the shore and the drops of dew, and to move the earth—if support is offered via Archimedes. Yet they are barbarous in their speech and poor in their writing, ignorant in their religion, perverse and corrupt, and noxious to politics, these obscure patriots who are unworthy of the hereditary vocation.[37]

It was not rare in these tumultuous years for the church to denounce scholars to the Turkish authorities as revolutionaries advocating the overthrow of the sultan. The patriarch Gregory V and the metropolitan of Chios, Plato, used this tactic against the director of the Chios Gymnasium, Neophytos Vamvas, as did the metropolitan of Smyrna against Constantine Economos, director of the Smyrna Gymnasium. The struggle against science reached its paroxysm in March 1821, when the Holy Synod was convened in Constantinople in order to put a stop to "philosophical" classes. The exact date was 23 March, after Christians had been arrested and executed following the rebellion of Prince Ypsilanti in Romania— but news of the Greek national uprising in Peloponnese had not yet reached the capital. Shortly afterward, on 10 April, the same Gregory V who had condemned scholars as subversive elements would himself be hanged, on the order of the sultan, because he had not been able to contain the rebellion. However, the victory of this rebellion seven years later would dramatically change the geography of Orthodoxy by dismantling the unifying Orthodox millet into several independent states, each with its own Orthodox church and distinct educational and scientific cultures.

CHAPTER THIRTEEN

The Scientific Modernization of an Orthodox State

Greece from Independence to the European Union

The rise of nationalism at the start of the nineteenth century upset the unity of the Orthodox Church and at the same time changed the cultural landscape for Christians of the Ottoman Empire. Each nation-state that emerged sought to establish its own autonomous church and its own educational structures in its own language. And so the Patriarchate of Constantinople lost the decisive role it had played for more than a thousand years, keeping only the title of *ecumenical*.[1]

The Greek Revolution of 1821 was the first European national revolution to result in the creation of a sovereign state. The leaders of this revolution had set themselves the goal of founding something modeled on their contemporary European nation-states. Born of the Enlightenment, the French Revolution, and also the romantic movement that engendered philo-Hellenism across Europe, the Greek Revolution succeeded thanks not only to European support secured by the philo-Hellenic movement but also to the geopolitical interests of the major powers. In 1828 Greek independence was imposed militarily by France, England, and Russia, which combined to achieve the "controlled" dismantling of the Ottoman Empire. However, the creation of a Greek national state stirred up nationalist revolts in other Christian nations of this empire, notably among the Serbs, who were not content with the level of autonomy that had recently been granted them. The perennial Balkan question, initially linked to Russia's ambition to govern the territories of Orthodox Slavs that formerly belonged to the Ottoman Empire, came onto the agenda. It was provisionally solved in 1878, when the Congress of Berlin, which brought together the European powers and the Ottoman Empire, recognized the independence of certain Balkan nations.

The creation of the Greek state in 1830—and, after the Congress of Berlin, the formal creation of Serbian, Romanian, and Montenegrin states and the recognition of Bulgaria—posed the fresh problem of frontiers that were not only political but cultural. Until then, the existence of a common political space (the Ottoman Empire), the authority of the patriarch of Constantinople as leader of the Orthodox millet, the absence of higher-educational structures in the Ottoman Empire, and the importance of Greek schools had all meant that such schools were the sole institution by which science was spread to the Orthodox people of southeastern Europe. The only exceptions were the territories controlled by Austria, whose Orthodox population could partake directly of Viennese education. After the restructuring of the political map based on nation-states, the Orthodox populations of these new countries wanted to study in new schools that taught in their national languages with the aid of national textbooks. Thus, the Greek language lost its status among these peoples.

Relations between science and religion became more complex, varying from nation to nation, each with its own autonomous national church. The first of these national churches was the Greek one, established in 1833 through the effort of a clerical partisan of the Enlightenment, Theoklitos Farmakides (1784–1860). Farmakides had studied at the Patriarchal School of Constantinople (which he found "authoritarian") and also at the University of Göttingen in Germany. He served as the adviser on ecclesiastical affairs to the regent of the young king of Greece, Otto of Bavaria; it was not in the king's interest to have the church of the new nation-state dependent on the patriarch, who was de facto an Ottoman subject—and who did not recognize the autonomy of the Greek church until 1850. The Bulgarian church followed in 1870, the Serbian in 1879, the Romanian in 1885, and the Albanian in 1922. These developments left the Patriarchate of Constantinople with an honorary primacy that no longer allowed it to interfere in local religious affairs.[2]

New State, New Structures, Old Problems

Following the principle of universal education proclaimed by the French Revolution, which was now spreading across Europe, the new Greek state created centralized education structures, organized and controlled by a specialized ministry—a concept that had not existed in the Ottoman Empire. From its foundation, this ministry combined education and religion; in fact, it was titled the Ministry of Education and Ecclesiastical Affairs. This conflation shows, on the one hand, the desire of the new nation-state to control the church and, on the other, the persistent concept that education could not be separated from Orthodox religion. Nevertheless, this ministry proved to be a crucial factor in the modernization of Greek society, organizing a national educational system comparable to many other European countries. The first educational system put in place, in the years 1834–37, comprised four levels: primary schools, lower secondary schools, gymnasia (or high schools), and the University of Athens. With the exception of the much smaller Ionian Academy, created in Corfu during the Napoleonic occupation and reorganized into a university during the period of the British protectorate, Athens was the first modern university established in the Balkans. At its creation in 1837, the University of Athens included faculties of theology, law, medicine, and philosophy. The sciences—physics, mathematics, and chemistry—belonged to the faculty of philosophy; an independent science faculty was not created until 1904. The first science professors of this university had all studied in leading European universities, which allowed them to integrate Greek higher education into the European scientific sphere.

The French Polytechnique played a very important role in the scientific and technological modernization of the nascent Greek nation-state. France used the Ecole Polytechnique (especially after 1816) for foreign policy purposes, by admitting students from countries that it wanted to aid or influence. As a result of this policy, Greek nationals became, along with Germans, one of the two most numerous foreign contingents until the end of the nineteenth century, when France's preference turned to Romania. The first Greek auditor (informal student) at the Polytechnique was Dorotheos Proios (1756–1821) in 1800, who later tried to modernize the Patriarchal School of Constantinople.[3] During the second Napoleonic occupation of the Ionian Islands (1807–11), the French government sent the Catholic mathematician and naval engineer Charles Dupin (1784–1873), who had studied at the Polytechnique with the mathematician Gaspar Monge (1746–1818) and become a protégé of the military engineer Lazare Carnot (1753–1823), to Corfu, where he founded the Ionian Academy. Dupin modeled the Ionian Academy after the Institute of Egypt, which meant providing a menu of cultural activities for members and "students" (candidates for membership) and organizing scientific lectures, courses, and scientific and literary competitions. After the departure of the French and during the British protectorate, the philo-Hellenic governor Lord Frederick North, 5th Earl of Guilford (1766–1827), tried to ensure the survival of the academy by transforming it into a university. He sent the mathematician John Carandino (1784–1835) to Paris, where he was admitted as a foreign auditor to the Polytechnique between 1820 and 1823. Upon his return to Corfu in 1824, Carandino helped to transform the academy into a university, with himself as director, Lord Guilford as president, and seven professors. Carandino was the first real mathematician of modern Greece. He wrote and translated from the French various treatises of higher mathematics. His students diffused French higher mathematics throughout the Ionian Islands and the Greek state in the nineteenth century. The Ionian Academy itself dissolved in 1864, the year these islands were integrated into the Greek state.[4]

The French Polytechnique also served as a model for the organization of other Greek institutions: a military school; a school of crafts and trades (including the fine arts); and an engineering school, a polytechnic that evolved into the National Technical University of Athens (called the Polytechnic). Throughout the nineteenth century, French policy and the vicissitudes of the Greek government determined the number and quality of Greek students at the French Polytechnique. Most of them came back to Greece to become university professors, politicians, engineers, company directors, or else military engineers (those who had been sent to Paris by the Greek military school). Among the fifty Greeks who went through the Ecole Polytechnique between 1829 and 1900, fifteen became professors in higher education (at the University of Athens, the Military School, the Polytechnic University), and six became heads of government ministries—one a prime minister.

The Greek engineers who studied at the Polytechnique or at other technical schools in France spearheaded the modernization of the new country, influencing not only the nation's material progress (railroads, highways, ports, mines, factories, urban infrastructures) but also its mental attitudes. Most of them were strongly influenced by the ideas of the socialist Henri de Saint-Simon (1760–1825) and later of the French hygienists and sanitary engineers, which they tried to apply in their own country. Thus, they fought for improvements in infrastructure, which often aroused the opposition of traditional Greek society. The story of water distribution in Athens is only one example of these conflicts: the construction of the running-water network encountered fierce opposition from small entrepreneurs who had traditionally brought in and sold water in the local neighborhoods.[5]

This spirit of modernization combined with Saint-Simonism had repercussions on the relations of scientists with the Orthodox Church. On the social level, Saint-Simonism proclaimed a fraternal society led by intellectuals, engineers, and manufacturers; on the metaphysical level, God was replaced by the universal law of gravitation. It is evident that no church had a place in that schema. Until the beginning of the nineteenth century, the links between scholars and the church was something rarely discussed, even in cases where scholars made innovations in education that were sometimes in contradiction with sacred texts. As we have seen, the new Greek nation conflated education and religion in the same ministry, thus perpetuating ties between science and religion. Engineers were the only men of science outside this network, having no obligations to the church. The Saint-Simonian spirit of these engineers, which privileged social progress over religious dogma, fostered their alienation from the traditional values of Orthodoxy.[6]

The cohabitation within the same ministry of education and religion—which continues to this day—did not, however, slow down the integration of Greece into the shared space of nineteenth-century European science. In cases of conflict between the two, as, for example, over the theory of evolution (see chapter 14), attacks on the modernizing university came not directly from the church but rather from parareligious groups that were struggling against the decline of old Orthodox values, which they feared would lead to the death of the homeland. Meanwhile, the university always kept up religious appearances, officially participating in Orthodox ceremonies and calling on the church to bless any inauguration. While maintaining this balance, professors at the University of Athens, the Polytechnic, and the Military School actively participated in mainstream science, publishing papers in specialized journals, generally French or German. As a rule, however, their research resulted from scientific collaborations with European colleagues; it was not the fruit of a national scientific community. Although capable of teaching science at an international level, the Greek scientific community was largely incapable of organizing national research.

The first establishment of the Greek nation-state that could be termed a research institute was the Observatory of Athens, made possible by a donation from Baron George Sinas (1783–1856), a very wealthy diaspora Greek who resided in Vienna. Sinas founded the observatory in 1842 with a view to strengthening Athens's prestige by means of a high-level scientific institution. In 1858 the German Johan Friedrich Julius Schmidt (1825–84) became director of the observatory. During his long tenure in Athens, Schmidt drew his famous map of the moon, the best one of the nineteenth century. He also organized seismic observations and, starting in 1861, edited the *Publications de l'Observatoire d'Athènes* (with the title appearing in French). Six years after his death, the reform-minded Greek government appointed a young researcher, Demetrius Eginitis (1862–1934), then in Paris, to direct the Observatory of Athens. A dynamic scientist, Eginitis eagerly sought to integrate Greek science into the European scientific scene by copying experimental methods, establishing ties, and patterning institutions on those abroad. Thanks to the donors he managed to mobilize, he was able to modernize the observatory by buying up-to-date scientific instruments and establishing meteorological and seismological services that, along with providing accurate time, made the establishment indispensable to Greek society. The increased prestige of the observatory reflected back on its director; having become the doyen of the philosophy faculty of the University of Athens, Eginitis managed to create an independent science faculty. Although Eginitis worked for the same goals as the Saint-Simonian engineers—the modernization of Greece—contrary to them, he declared his faith in Orthodoxy. Although not a partisan of a six-day creation, he believed in a superior nonmaterial force that had created matter and given it energy. His beliefs coincided with those of the vast majority of the Greek scholars of the turn of the century: that there was continuity in Greece from the antiquity to contemporary times, integrating in the same cultural space Hellenic philosophy and Christianity throughout the Byzantine period. Eginitis contributed to this idea by work aiming to demonstrate the stability of the Greek climate from antiquity to the modern era.[7]

Greek chemists, led by the Vienna-born, German-trained Anastasios Christomanos (1841–1906), who began teaching at the University of Athens in 1866, were among the next Greek scientists to rise to international status, thanks to the needs of the nascent chemical industry. In physics, the first up-to-standard experimental laboratory—organized at the University of Athens by Timoleon Argyropoulos (1847–1912), who had studied at the Sorbonne—did not appear until the last decade of the nineteenth century. Concerning science and religion, Argyropoulos, claimed, against the materialists, the existence of a superior spiritual force. As he believed that everything is movement, he considered that God is the prime mover of the world.[8]

The delayed development of scientific disciplines in Greece shows the relative weakness of science in relation to letters, both in the university and in the collective consciousness of the young state. In effect, the philo-Hellenic movement and the romantic movement had forged the idea that Greece's strength rested on its glorious past and, hence, that it was more profitable to study this past than to develop science. Moreover, a significant part of the Orthodox Church perpetuated phobic behavior toward the West because, like the old patriarchates, it regarded the teaching of science as equivalent to teaching Western culture, thereby distancing students from Orthodoxy. As a consequence, the social sciences and humanities rather than the natural sciences dominated the University of Athens. The creation of an independent science faculty was an important step toward overturning these power relations among disciplines, as it was in other universities in western Europe. Women remained almost totally absent from Greek science until the twentieth century. The first female to obtain a diploma in engineering from the Polytechnic University did so in 1923.[9]

The Greek church's fear of the West, especially Catholicism, meant that it continued the tradition of the Patriarchate of Constantinople in fiercely opposing the introduction of the Gregorian calendar (considered as papist) into Greece. The patriarchs Anthimos VII and Ioakeim III broached the topic of calendar reform at the turn of the century, but they ultimately dismissed it. But after the adoption of the new calendar by Soviet Russia in 1918 and by Romania in 1919, the problem came back in force in the Greek political scene, for Greece was now the last Christian European country not to have adopted the Gregorian calendar. Eginitis, who had worked in 1916 to standardize time by integrating Greece into the European system of time zones, exploited the political situation created by the military coup d'état of 1922 and convinced the new government to adopt the Gregorian calendar on 17 February 1923, which thereby became 1 March. This caught the church by surprise, but in 1924 it reluctantly adopted the new calendar, with the exception of the date of Easter, which has remained down to the present calculated according to the Julian calendar. However, a few isles of resistance remained, most notably the Autonomous Monastic State of the Holy Mountain, Athos, which still follows the Julian tradition.

The war between Greece and Turkey that followed the First World War marked the end of the territorial expansion of the Greek state (with the exception of Dodecanese, the islands in the southeast of the Aegean Sea, which were annexed at the end of the Second World War). Before the First World War, the aspiration to create a Greater Greece that would include Asia Minor and Constantinople had postponed plans to found universities other than the one in Athens. Indeed, a Greek university, the University of Ionia, had been founded in Asia Minor, at Smyrna, during the occupation of the city by the Greek army in 1919, but this university never got off the ground because Smyrna was soon reoccupied by Turkish troops. Thus, it was not until after the defeat in Asia Minor and the consequent failure of the University of Ionia that Greece founded, in 1925, its second university (or third, if one counts the Polytechnic in Athens), this time inside its recognized borders, in Thessalonica. A year later, Demetrius Eginitis wrote the act founding the Academy of Athens, called the Academy of Sciences, Letters and Fine Arts, whose fine neoclassical building had existed since 1887, thanks to its donation by Baron Simon Sinas. Eginitis took advantage of the fact that he had just been named minister of national education and ecclesiastical affairs by the government of Theodoros Pangalos, the outcome of a military coup d'état. It took the prestige of Eginitis and the strong-arm democracy of Pangalos to resolve the problem of who could become members of this venerable institution.

The Cold War and Scientific Research

The turbulent years that followed the foundation of both the University of Thessalonica and the Academy of Athens, and political vicissitudes associated with the civil war of 1946–49, pitting British- and American-backed government forces against the communist-controlled Democratic Army of Greece, inhibited the founding of new scientific institutions like those springing up in other European countries during these years. Until the 1950s, the three universities (Athens, the Polytechnic University, and Thessalonica), plus the higher-education School of Agriculture, founded in 1920, remained primarily institutions for teaching science, while the Athens Academy was a society of recognized scholars. Meanwhile, the Observatory of Athens was getting by with equipment acquired at the start of the century, and scientists at the University of Athens were using old instruments dumped by the Germans after the First World War as part of their reparations imposed by the Allies. After the Second World War, Greece received the highest per capita funding under the Marshall Plan, but this money went largely for military purposes because of the ongoing civil war. Apart from the problem of outdated equipment, the exclusion of the Left from politics during the right-wing dictatorship of Ioannis Metaxas (from 1936 to 1941) had negative repercussions on the universities and the academy. This exclusion continued after the war because of the communist insurrection. Metaxas's regime's slogan, "work, religion and family," perpetuated by conservative regimes after the end of the Second World War, did not favor radical attitudes. Meanwhile, Orthodox scientists were organizing a network aiming to check the religious affiliation of scientists applying for posts at state institutions. As left-wing scientists saw their careers compromised, some of them sought their fortunes abroad.

This situation changed around the middle of the 1950s during the Cold War. As one of the countries bordering the communist bloc, Greece became strategic to the West. When the communist countries began organizing scientific research in institutes or science academies, the Greek government responded, with the help of the United States, by creating its own national research institutions. In 1958 the U.S. National Institutes of Health invited Leonidas Zervas (1902–80), a professor of chemistry at the University of Athens and member of the Athens Academy who as a young researcher had spent several years at the Rockefeller Institute for Medical Research, to its laboratories in Bethesda, Maryland. There he laid plans with Americans to found a research center in Greece. Afterward, he authored a report about his meetings with the Americans:

The higher officials of the great foundations like the Rockefeller Foundation etc., and obviously Mr. [Waldemar] Nielsen, are extremely capable, educated, and fully aware of the situation in various countries from the social, scientific, and economic standpoints, and they collaborate closely with the State Department. For some time, these large foundations—and by all accounts official America—have begun to understand that finding solutions for the free world to survive did not mean recalling or "retaining" as large a number as possible of our scientists who are so highly reputed in the U.S., but rather that it was imperative to furnish aid for the improvement and creation of local science in free countries.[10]

In July 1958 Zervas met Ide Carter, chief of the Middle Eastern Division—Greece, Iran, and Turkey—of the International Cooperation Administration (precursor of the United States Agency for International Development). Carter gave his approval for 150 million drachmas (five million dollars) of unused Marshal Plan funds to be released to finance the Royal Research Foundation (which would become the National Hellenic Research Foundation after the abolition of the monarchy in 1974). The Ford Foundation offered an additional quarter-million dollars. According to Ioannis Pesmazoglou, one of the project's instigators, the plan was to create a research center in the image of the French National Center for Scientific Research (CNRS) or Germany's Kaiser Wilhelm Institute.[11] Apart from organizing research, the goal of the Greek research center was to develop the ethos of a scientific community, elevating scientific excellence over the then-current client-driven system and guaranteeing independence and freedom for researchers. As Zervas wrote of the United States,

> Here and in Western Europe and Russia, people agreed long ago about what science is, whether pure or applied, and what scientific research and art both are. You know, perhaps better than me, that in our country all of those things are not properly defined, and are instead used by interested parties for various and sometimes incredible purposes. . . . For God's sake, avoid giving the impression that the Royal Foundation is trying to impose censorship or to appropriate the work of a researcher! Intellectual property is conserved even in Russia![12]

In 1955, at the Palace of the United Nations in Geneva, a major international conference took place on the peaceful uses of nuclear energy. This conference was a product of President Dwight D. Eisenhower's atomic policy aimed at scientific collaboration and the supervised export of American technology. At the same time, Athens hosted an exhibition under the auspices of the king of Greece and in cooperation with the Central Intelligence Agency, on the peaceful use of atomic power, where the U.S. ambassador, Cavendish Cannon, and the president and secretary-general of the Hellenic Agency for Nuclear Energy (Athanasios Sapnidis and Theodore Kouyoumzelis) vaunted the benefits of nuclear energy. The Hellenic Agency for Nuclear Energy had been created only the year before, following Greece's joining CERN (the European Organization for Nuclear Research) in 1952. As in the case of other countries in the American sphere of influence, an agreement had been signed between Greece and the United States for the construction of a nuclear reactor and its provision with uranium, which would be installed in a specialized research center. Thus was born the Demokritos Center for Nuclear Research (after Democritus, the ancient Greek philosopher of atomic theory), whose reactor was inaugurated in 1961. Apart from the construction and maintenance of the reactor, the United States offered $350,000 to Demokritos; it and the Royal Research Foundation were placed under the auspices of King Paul (who was also the honorary president of the foundation's board). Along with Queen Frederica, he actively promoted the idea of science for peace. In this era, the royal couple organized physics seminars for the Athenian public on the historic hill of Pnyx, to which were invited celebrated physicists such as Werner Heisenberg, who had played an important role in the creation of CERN, and the 1949 Nobel Laureate for the discovery of mesons, Hideki Yukawa.[13]

The fact that the research centers were from the outset placed under the auspices of the palace spared them criticism from conservative circles and allowed them to hire researchers independently of candidates' political and religious beliefs. This was achieved at a time, after the civil war, when Greece was playing a pivotal position during the Cold War, when it was almost impossible for scientists who declared themselves as "on the left" or "atheist" to be hired by the universities. Although the Orthodox Church feared the effects of some science teaching, especially Darwinist ideas (see chapter 14), it did not oppose scientific research itself; it thus largely ignored the new research centers.

The establishment of research centers at the end of the 1950s was an important step toward the integration of Greece into the world of international scientific research. However, this process was compromised during the years from 1967 to 1974, because, under the régime of the colonels, several scientists preferred to pursue careers abroad, either in western Europe or in the United States. The reestablishment of democracy in 1974 marked the beginning of a scientific revival, which was accelerated by the return to the country of a number of researchers who had meanwhile formed links with foreign laboratories, and also by the adherence of Greece to the European Union in 1981, as the tenth member-state, thanks to the reforming efforts of Prime Minister Constantine Karamanlis (1907–98).

The ties of the Greek state with the Orthodox Church were, at its creation in 1830, indisputable. State ideology taught in schools proclaimed that the Greek nation had survived four centuries of Ottoman domination because of the Orthodox Church. The creation of a ministry combining education and religion was the institutional translation of this statement. Despite the pressure put by Greek intellectuals on the government officially to separate education and religion, this has not yet happened, because of the popularity of the Greek Orthodox Church. Nevertheless, the involvement of this church with higher education and scientific institutions has been unobtrusive. The other side of the new Greek state ideology, that of modernization, prevailed over conservative Orthodox behavior, even during conservative regimes that favored the involvement of the church in social life, such as during Metaxas's dictatorship or the right-wing regimes of the first two decades after the Second World War.

CHAPTER FOURTEEN

Science and Religion in the Greek State

Materialism and Darwinism

Between 1804 and 1827, no fewer than six Greek students audited the lectures offered by the notorious evolutionist Jean-Baptiste Lamarck (1744–1829), professor of zoology at the Museum of Natural History in Paris. However, none of them returned home to spread the French naturalist's theory. Thus, it was left to Alexander Theotokis (1822–1904), a student of Henri Ducrotay de Blainville's (1777–1850) at the Paris museum, to first introduce Greeks to the idea of organic evolution, in his book *General Zoological Tables; or, Forerunner to Greek Zoology,* published in 1848.[1] Theotokis, who belonged to a well-known family from Corfu, presented three different schools of thought: immutability of species; immutability of species for the most part, but with possible changes in secondary characteristics; and change based on the theories of Lamarck. Following Blainville, Theotokis preferred the second school and accepted that God created species. Perhaps because he never held an academic post or any other position of influence, Theotokis's ideas did not have a significant impact in Greece.[2]

The first mention of the English naturalist Charles Darwin appeared in the *Attic Calendar* of 1869, a sort of almanac with articles of general topicality. In the entry "The Spoils of Pikermi," the anonymous author wrote: "Apart from the great number of [fossil] animals discovered, the collection found in Pikermi gives force and scientific significance to the theory of Darwin on zoogeny."[3] In the years from 1855 to 1860, the French paleontologist Albert Gaudry had conducted excavations in fossil deposits dating from the Miocene at Pikermi in Attica. Darwin was aware of Gaudry's research and mentioned the "discoveries of Attica" in both the *Origin of Species* (second edition of 1872) and *The Descent of Man* (1871).[4] Gaudry's excavations and the remains of strange animals found there had created a stir in Athenian society; so it is not surprising that the author of the *Attic Calendar* knew about Darwin's theory and the relation of the Pikermi discoveries to this theory even before Darwin mentioned them.[5] As professor at the University of Athens, Hercules Mitsopoulos had taken part in these digs in 1851, before the arrival of Gaudry, and had highlighted the "antediluvian" animals. The author of the article in the *Attic Calendar* also briefly and favorably presented Darwin's theory on the "alteration of beings."[6]

A few years later, in 1873, an assistant professor of chemistry at the University of Athens, Leandros Dosios, introduced evolution to the educated Athenian public in a lecture at the Parnassos Philological Society. We do not know if there was an immediate reaction to Dosios's lecture, titled "The Struggle for Existence," but three years later the first serious attack against evolution came from an assistant professor of the faculty of theology at the University of Athens, Spyridon Sougras, who wrote *The Most Recent Phase of Materialism: Darwinism and Its Lack of Foundation*, a polemical pamphlet against the threat of atheistic materialism.[7] "From the start, the question of evolution or changes in animals and vegetables has attracted the attention of all scientists, because of its imbecilic and horrible results, meaning the common origin of man and the monkey," wrote Sougras. "It also attracts religious interest, for theology ought also to be concerned with this question, since by its simple suppositions alone, it risks inducing error even among those who are totally ignorant of this theory."[8] Sougras strongly feared that Darwinian theory would push the Greeks toward socialism and communism, and in the sequel to his book he attacked Karl Marx and the German political activist Ferdinand Lasalle (1825–64). Sougras's arguments were both philosophical and theological and were based on two intellectual qualities, morality and religious sentiment.

In 1877, one year after Sougras's book appeared, the professor of botany at the University of Athens, Spyridon Miliarakis, published a serialized biography of Darwin in the learned journal *Estia*, a biography issued as a small book in 1880.[9] In the preface, Miliarakis criticized those who "make judgments and improvise discourses on everything, without ever studying the subject about which they are expressing conflicting opinions, as is the case with those who are opposed to the theories of Darwin." Although a convinced Darwinian, Miliarakis proposed calling it a theory of "development" instead of "evolution": "We think the word *development* best agrees with the theory of Darwinism, since it carries the notion of the perfectibility toward which any organic nature constantly tends." In general, with the exception of the faculty of theology, the academics at the University of Athens considered Darwinian theory rather positively; the few criticisms that did appear were typically moderate. One of the most negative remarks came from Ioannis Soutsos, professor of political economy and sometime rector of the university, who in 1877 criticized Darwinism for emphasizing the struggle for existence and hence promoting violence.[10]

Darwin, Haeckel, and the "Ethnocide Group"

The largely lay defense of creationism by the Orthodox Church erupted at the end of the nineteenth century as part of the political struggle between the partisans of modernization and reform—those of the reformist party of Charilaos Trikoupis and the populist forces of the national party, led by Theodore Deligiannis. Ioannis Skaltsounis (1821–1905), who had studied law at the Ionian Academy before going to Italy to study in Siena and Pisa, was a respected businessman and lawyer. After 1873 he went to Trieste, where a community of Greek merchants was flourishing. In addition to conducting his regular business, Skaltsounis wrote books and articles on materialism and the origin of humans.[11] According to Skaltsounis, the most dangerous materialist was not Darwin but the German biologist Ernst Haeckel (1834–1919). "Darwin proposed as probable the existence of an archetype for the origin of man and the monkey, and he several times admitted the logical weakness of his explanations," observed Skaltsounis. "But Haeckel tried to define with mathematical precision the bestial origin of man."[12] Skaltsounis argued that the University of Athens, with its impious professors, should be abolished. This problem, as he saw it, went beyond the teaching of evolutionism; he feared that prestigious university professors might legitimize materialist ideas within Greek Orthodox society. To harmonize science and religion, he maintained that a supernatural force had created all living beings, a claim that flew in the face of evolutionary biology.[13]

Theodore Tsikopoulos, a doctor who had studied in Germany and worked as a physician on the construction of the railway from Thessalonica to Istanbul, answered Skaltsounis in a book published in 1894.[14] Tsikopoulos dedicated his work to the politician who was fighting for the modernization of Greece, Charilaos Trikoupis. The physician's work had put him in contact with engineers who were building the railroad, and these men were the principal vectors of modernization in both Greece and the Ottoman Empire. According to Tsikopoulos, the dilemma of choosing between self-engendering and creationism did not arise: life must be the combination of force (confused with energy) and matter, because the triad of life-force-matter forms a unit. Manifestly, Tsikopoulos was influenced by the monism of Haeckel, whose ideas had already spread in Greece. The debate pitted materialism against idealism, and evolutionism against antievolutionism. Tsikopoulos, as a proponent of modernization, feared any interference by the church in scientific affairs would lead to the humiliation and ruin of modern Greece.

A year after Tsikopoulos's answer to the anti-Darwinians, Philopoimen Stephanides, a philology student, gave a funeral oration for his comrade Vasileios Katsanis, who had just committed suicide. This speech was published under the title "The Hypotheses of Materialism and Darwinism and the Greek University."[15] Stephanides thought that the professors at the University of Athens had poisoned the heart of his young friend by their Darwinian teaching of evolution and had thereby brought about his suicide. He closed his oration by calling for the death penalty for those who publicly questioned the existence of God and the soul, on the grounds that they were guilty of supreme treason to Greece.

The Darwinian Greeks—a disparate group of professors, physicians, and other intellectuals—found an important and unexpected ally in the person of the German philosopher and theologian Eduard Zeller (1814–1908), who had discovered that some ancient Greeks, notably the pre-Socratics, had in fact anticipated Darwinism. That was all it took for the Darwinians to add national pride to the reasons for accepting evolution. They translated Zeller's writings into Greek to demonstrate that Darwinism descended from the philosophy of the ancients.[16]

The Orthodox Church did not formally condemn the theory of evolution. Condemnations in the form of pamphlets were primarily the work of reactionary social circles, which were often affiliated with the Populist Party. These circles feared the changes that the promotion of modernization was likely to bring, on both the infrastructural and the institutional levels, which would have consequences on ways of thinking. Their prime ideological enemy was not Darwinism but materialism; however, they saw evolution as a tool of the materialists at the University of Athens.

Despite this opposition, interest in evolution soared at the university. In 1880 a course on evolution theory given by Ioannis Zochios (1840–1912), professor of physiology at the faculty of medicine, was so successful that no university hall was spacious enough to hold the audience. His success led the Holy Synod of the Church of Greece to put pressure on the university to put a stop to this unacceptable situation. Greek Orthodox churchmen feared the theological implications of Darwin's theory and absolutely rejected the common origin of apes and men, as defended by Haeckel and Darwin. For the church, the six-day creation had not been open for discussion. The University Senate discussed the affair at its meeting of 4 April 1880. According to the vice-rector and the doyen of the faculty of theology, Zochios "in his classes on Darwinism was led to conclusions that were offensive to the dominant religion."[17] Despite this rebuke, the council failed to take action against Zochios, and in the end the rector simply recommended that Zochios be more prudent when discussing ideas that might offend the religious sentiments of his audience.

By the end of the nineteenth century, the Greek church was becoming synonymous with conservatism. It lashed out at the materialist ideas associated with the new political forces of socialism and communism, which (according to the church) were poisoning youth and luring people away from the religion of their ancestors. The reaction against materialism paralleled the church's response at the start of the nineteenth century to the radical ideas associated with the French Revolution. In the current situation, Orthodox leaders saw revolution and socialism as manifestations of the Western atheism that had been introduced into Greece by science education. The Holy Synod feared that Darwinian ideas would furnish arguments to materialists. Because these ideas were either taught at the University of Athens or promoted by its professors, it was against this establishment that ecclesiastical arrows were shot.

Given the institutional imbrication of church and state that made the Greek church dependent on the government, the Holy Synod was always careful not to make decisions that might offend state institutions. Thus, the role of attacking the University of Athens was handled by publications close to the church but not officially linked to the Holy Synod. In 1887 the journal *Anaplasis* (*New Creation*), published by an association of the same name, appeared. With an impressive circulation of 4,500, it became an important conservative voice; the Holy Synod recommended it to the clergy, which in turn recommended it to believers.[18] The Anaplasis Association supported the church in its new struggle against materialism, as clearly stipulated in its founding manifesto:

> The Association is composed of more than two thousand members of the civilized world; among them are the distinguished patriarchs of our Orthodox Church, the metropolitans of the Orthodox states, Greek and foreign scientists, celebrated theologians. . . . It has the principal goal of safeguarding with all its strength our holy religion that is offered to us by God and our morality against all the enemies of Christianity that have reappeared, above all against the pernicious materialism that has been rife for some time, and which, in the name of a new erroneous science, denies any spiritual existence and openly combats the great truths of Christianity.[19]

Occupying the opposite end of the ideological spectrum from *Anaplasis* was the journal *Prometheus*, published between 1890 and 1892 by the professor of geology at the University of Athens, Constantine Mitsopoulos (1844–1911), in collaboration with two others, Nicolas Germanos (1864–1932) and Stamatios Valvis, assistant professor to the former. The editors stated their goal in the subtitle: *The Diffusion of Exact and Applied Sciences.* The journal, presenting the ideas of Darwin and Haeckel in a rigorous manner and publishing articles written by Greek scientists as well as several translations, enjoyed great success, especially among students at the University of Athens. *Prometheus* also published philosophical essays in defense of science, such as a translation of Charles-Auguste Mallet's *History of Ionian Philosophy*, which defended scientific logic against theology. *Prometheus* became a forum for advocating deistic and materialist propositions in the ideological context of scientific positivism at the turn of the century. The fact that the journal was published by three academics and that most of the articles and reviews in defense of Darwin and Haeckel were written by one of them, Stamatios Valvis, while he was serving as an assistant professor at the University of Athens, did not improve relations between the Greek church and this venerable institution. The anonymous editor of the church-related *Anaplasis* wrote in despair:

> Amidst the moral paralysis that is devastating the country, professors at the National University have judged the moment propitious to teach the Greek public through the press that the convictions about the existence of a creating spiritual force are devoid of any scientific foundation, for it is experimentally demonstrated that there exists nothing outside matter and that all forces and capacities of man flow from the vibrations of the body. Such is materialism, according to the editors of *Prometheus*, where reality and truth are found.[20]

The conflict reached its height during the decade of the 1890s. The Darwinians sometimes advanced extreme positions—that everything could be explained by matter, for example—which exasperated the Orthodox, who counterattacked in articles in *Anaplasis*, calling for measures to be taken against materialist education. One contributor urged the Holy Synod to intervene against the "ethnocidal group" that was "poisoning the nation." "The Church should show it is alive," the author declared. "The state should throw out of the University and *gymnasia* any teacher of materialist theories."[21] Despite such extreme views, the debate remained rhetorical. There were no repercussions for the teachers because the government refused to intervene.

After a fifteen-year hiatus, the debate over monkeys-to-man broke out again. The occasion was an educational experiment in the Greek schools conducted by

Professor Alexander Delmouzos (1880–1954). After having studied pedagogy at Jena, Delmouzos was named in 1908 director of the Girls' School in Volos, where he introduced instruction in vernacular Greek and explored ways of developing the students' critical thinking. Instruction in the vernacular had sparked an ongoing debate since the formation of the Greek state: should Greece adopt the purest language in order to revive the ancient heritage or the spoken (demotic) language that made official documents and education more comprehensible? Delmouzos's novelties in education disturbed conservatives, notably the metropolitan Germanos Mavromatis, who tried to get rid of Delmouzos and his collaborators, whom he labeled as dangerous anarchists. Orthodox believers organized protests, which degenerated to such a point that the cavalry was called in to defend the school and Delmouzos's house. In the face of this strong popular opposition, the Volos municipal council decided to close the school in 1911 and to prosecute Delmouzos and his collaborators. The trial of the "atheists," as it was called, took place in Nafplion in 1914. The reformers were accused of proselytizing for atheism. To support this accusation, the prosecution ironically invoked the theory of evolution: "In various periods, from September 1908 until the end of March 1911, in Volos and in Larissa, principally at the Workers' Foundation and the School for Girls in Volos, they [the accused], teaching out loud or with the aid of printed brochures, tried to proselytize in favor of a so-called religious dogma, i.e., atheism. These actions are incompatible with the preservation of the political order, for they teach that God does not exist . . . that man was created from monkeys, that God is a cucumber, etc."

The court acquitted the accused for lack of proof. Although the discussion of the origin of man had constituted only one aspect of the trial, the defenders of Darwinism noted that this was the first time in Greece that teachers had been prosecuted for teaching the theory of evolution. Neither the government nor the Holy Synod of the Church of Greece was directly involved in this affair; nevertheless, the trial came to symbolize their struggle against the modernization of education.

In the years that followed, the modernizing forces and the church's conservative circles reached a standoff: evolution would not be taught in Greek secondary schools. Twelve years after the trial, in 1926, the evolutionists achieved a minor victory when the newly founded Academy of Athens elected as two of its first members Rigas Nicolaidis (1856–1928) and Georgios Sklavounos (1899–1938), both of whom had taught evolution at the university. The fact that the primary founder of this venerable Academy, the distinguished astronomer Demetrius

Eginitis, was a prominent member of the Orthodox Church shows the eagerness of moderates on both sides of the debate to work for reconciliation. Eginitis viewed the evolutionists primarily as his scientific colleagues.

At times during the first quarter of the twentieth century, both partisans and enemies of materialism were sometimes confused with partisans and enemies of the monarchy. For example, in 1917 the prime minister authorized the firing of some royalists at the Polytechnic University of Athens (today, the National Technical University of Athens), including Aristippos Kousidis, a professor of applied mechanics. When King Constantine returned to the throne in 1920, Kousidis was rehired. He later became director of Greece's Post-Telegraph-Telephone Department and adviser to the government of Ethiopia. This moderate and deeply Christian engineer wrote a small book expressing the ideas of Greek Christian scientists, *Religion, Science, Society* (1935). Kousidis tried to harmonize the latest scientific discoveries with Orthodox Christianity. According to Kousidis, Einstein's famous equation $E = mc^2$ shows that science refutes materialism, since matter is only energy, similar to a spiritual force. But it was Darwinism that preoccupied Kousidis. Although he supported scientific dating and accepted the evidence of human antiquity—perhaps older than some species of monkeys—he believed that God created species from time to time, following a design that he alone knows. The spontaneous appearance of such divinely created species would make the theory of evolution unnecessary.[22]

Christians and Communists

The propagation of atheistic socialism strongly marked academia after the first quarter of the twentieth century. The dominant tendency among Greek intellectuals went in that direction, with Christian scientists remaining a minority. The one scientific institution where Christianity flourished was the National Observatory of Athens. The Christian Stavros Plakidis (1893–1990), who became director of the Observatory's Astronomical Institute in 1935, and his successor Demetrios Kotsakis (1909–86) both followed a tradition that was sympathetic to Christian values. Kotsakis was one of the rare Greek scientists of renown who was also an active member of a Christian fraternal organization, the influential Fraternity of Theologians of Zoë (life), founded in 1907.

In 1937 the Fraternity of Theologians of Zoë created the Christian Union of Scientists, which published the journal *Aktines* (*Rays*). On Christmas Day 1946, the year the Greek Civil War broke out, the organization issued the *Declaration of the Christian Union of Scientists*, a 180-page pamphlet addressed to "all Greeks" and signed by 222 non-Marxists in the sciences and letters, most of them university professors. Remarkably, some of the signatories had been in the past attacked by the church as enemies of Orthodoxy. The bipartisan political atmosphere of the era, probably combined with pressure from the government, enabled the declaration to attain its goal of giving voice to non-Marxist scientists. The declaration's most striking feature was its political prudence at a time of civil war. It contained no explicit condemnation of communism, though it denounced materialism, Darwinism, monism, and the teaching of Freudianism. Mostly, it advocated the value of spirituality and work. Unlike the Orthodox traditionalists who had previously set the tone within the Greek church, the Christian Union of Scientists emphasized the importance of science for society, while noting that "science is everywhere discouraged and the scientist has long ceased to be a leader in society and to be honored as such."[23] The *Declaration* blamed the spiritual decadence of the age not on material poverty but on spiritual poverty. Maintaining that Christianity should serve as "the foundation of our modern society," it lashed out at those who considered atheism not "as a singular deviation but as the foundation of civilization." Previously, it noted, "*atheistic* was synonymous with crazy, eccentric or villainous," whereas belief in a Creator was the foundation of any philosophical thought.[24] The *Declaration* pointed out that, since independence, Greece, instead of basing its civilization on its Christian tradition (which had been the response of Hellenism to the Roman yoke), had drifted away by imitating everything that came from the West.

The *Declaration* aimed to reconcile science and religion by citing recent "discoveries" that, according to the authors (who remained anonymous, although Kotsakis probably played a main role), were not in contradiction with Christianity. These included the creation of the world by a Supreme Being that preexisted it and will exist eternally, the existence of a spiritual world that does not obey the laws of matter (where humans participate through their souls), and the revelation of God by Christ, as well as all the doctrines that flow from this. Because none of these beliefs can be falsified by scientific evidence, they remain securely in the realm of faith.

In defending the Orthodox faith, the *Declaration* took on not only Haeckel but also the materialist philosopher and physiologist Ludwig Büchner (1824–99), whose book *Force and Matter* (1855) had been translated from German into Greek in 1910 and enjoyed several editions. The *Declaration* charged these men with trying to replace faith with science—and not even science that was demonstrated by empirical methods. True science, it argued, did not teach that living organisms could spring from inert matter. Thus, Büchner and Haeckel were wrong to believe that Darwinism provided a scientific basis for the theory of autogenesis:

> We should at first notice that Darwin's theory in its pure and original aspect did not have any of the meaning that was later attributed to it, meaning the automatic creation by chance of the world and especially of man, with no divine creative force. [Darwin] as a scientist, knew how to limit his imagination. Nevertheless, others like [Thomas] Huxley in England and [Karl] Vogt and Haeckel in Germany, who did not actually do scientific work but propagandized it through popular editions, utilized this theory for their own purposes. At that time, the effort at atheist propaganda in the name of science was at its height.[25]

The aspersions on Huxley, Vogt, and Haeckel were patently untrue, but then the primary goal of the *Declaration* was to show that science refuted evolution. The latest scientific findings, it falsely insisted, showed the stability of species, not their variability.[26]

In discussing the origin of man, the *Declaration* distinguished between Darwin and the popularization of his work by the "atheists," who used the existence of *pithecanthropus* (a partial hominid skeleton discovered in Java in the early 1890s) as proof of the filiation of our species to the apes. Arguing against the possibility of this fossil being prehuman, it suggested that paleontological findings give evidence only of degenerate human races. In any case, it went on, "the creation of the world by God would certainly not be undermined by the existence of any species of pithecoïds. Man's own spirituality is accentuated and exalted by the existence of animals that, despite having bodies that resembles man's, have no spiritual life." To support its claims, the *Declaration* frequently cited the Swedish geneticist Nils Heribert-Nilsson (1883–1955), one of the few well-known scientists to doubt evolution.[27]

According to the *Declaration*, the physical sciences cannot give answers to questions that are not about material nature. Moreover, the sciences teach us that this material nature and the laws that govern it are not sufficient to explain all of reality, notably human life. At the point where science stops, religion takes over. Miracles do not contradict science; they merely demonstrate the existence of another, superior force that manifests its own energy. Scientists rightly claim not to have encountered this supernatural force in their observations, for it does not belong to their fields of research.[28]

Although the *Declaration* makes no direct allusion to communism, it was nevertheless a political response by Orthodox Christians to the materialist ideas propagated by the communist camp, which at the time was fighting hard for power. The civil war and the strong-arm republic that followed it dug a chasm between "Left" and "Right" that lasted for almost thirty years. Complicating the situation was the fact that some committed Christian Darwinists, such as Vasileios (Vasos) Krimbas (1889–1965), a professor at the Agricultural University of Athens, signed the *Declaration*. In 1950 the fundamentalist Orthodox priest Avgoustinos Kantiotis (1907–2010) accused Krimbas of being an enemy of Orthodoxy and depicted him with a tail and pointed ears signing the *Declaration*.[29] Krimbas was elected to the Academy in 1960, and Kantiotis became metropolitan of Florina seven years later.

The *Declaration* had no effect on the teaching of Darwinism in the faculties of science. Nevertheless, the faculties of theology of the University of Athens and of Thessalonica (which functioned after 1945) continued to take an anti-Darwinian line. In 1969, during the dictatorship of the colonels after the coup d'état of 1967, the Russian-born Theodosius Grigorovich Dobzhansky (1900–75), a key figure in constructing the modern evolutionary synthesis, was invited to a conference of the Greek Anthropological Society. This famous geneticist, who had remained Orthodox, was wounded by the attack he suffered at the hands of Greek theologians, especially Marcos Siotis of the University of Athens, who protested against the idea that mankind might have descended from brute ancestors because that would contradict the book of Genesis. Dobzhansky protested as both a scientist and a member of the Eastern Orthodox Church that he was supporting humans "being created in God's image by means of evolutionary development." After the event, he observed, "Fortunately, the hidebound rigidity of the Greek section is not shared by the Eastern Church as a whole."[30]

In 1984 conservative members of the Greek Orthodox Church renewed their battle against evolution, shortly after the Ministry of National Education and Ecclesiastical Affairs introduced into secondary education a textbook by Lefteris Stavrianos, *The History of the Human Race*. The fundamentalist Kantiotis fought ferociously against this:

Do you know the conclusions of famous scientists, biologists, geneticists, embryologists, geologists, paleontologists, at the conferences in Chicago in 1980 and in Liverpool in 1982? If you don't, then we inform you that these men of science concluded that the origin of man is not yet demonstrated in a scientific manner. The theory of evolution was condemned, and the various genealogical trees, born of the imaginations of those who unreservedly support it, should disappear from the education books.[31]

Critics also accused Stavrianos's book of underestimating the role of Hellenism in the history of humanity. Under attack, the ministry finally withdrew it as an approved textbook in 1990.

By this time, positions were hardening within the church. A quarter century earlier, Kallinikos Karoussos (born Konstantinos,1926–2008), later elected metropolitan of Piraeus, had collaborated with Christos Paraskevaidis (1939–2008), the future archbishop of Greece known as Christodoulos, in founding the Christian fraternity *Chrysopigi*. This aggressively fundamentalist fraternity fought against the teaching of evolution, while confusing evolution with Marxism:

> Marxism is not only an economic, social, and political system, it is a materialist and atheistic theory; its basis is historical materialism and that is why it is contrary to the Christian idea of life. In our day, the materialists strike mercilessly at anything that is related to God and the Church. And they forge myths, like that of the origin of man from the monkey, which they propagate by any means possible. This theory, although it is not scientifically demonstrated and has been rejected by all serious scientists, is nevertheless taught by our compatriots the materialists in schools, thus poisoning our trusting and defenseless children.[32]

Kallinikos went on to found the Piraeus association of scientists in 1993, dedicated to fostering nationalist and antievolutionary ideas. Christodoulos, who ascended the archbishop's throne in 1998, enjoyed great popular support while continuing to promote the fundamentalist line advocated by *Chrysopigi*. As archbishop, he soon found himself at loggerheads with the Ministry of National Education and Ecclesiastical Affairs, something that previous archbishops had tried to avoid. His premature death in 2008 and his succession by the moderate Ieronimos II (b. 1938) spelled a switch to noninterference by the Orthodox Church of Greece in matters of education and the sciences.

In a widely read survey of the public acceptance of evolution in thirty-four European countries (plus Japan and the United Sates) published in 2006, Greece ranked seventh from the bottom, with just over 50 percent of those Greeks questioned accepting evolution. The four primarily Orthodox countries in the survey—Greece, Cyprus, Bulgaria, and Romania—were clustered in the bottom eight. Cyprus surpassed only the United States and Turkey. Once the center of Orthodoxy but now largely Muslim, Turkey came in dead last. A survey of Eastern Orthodox believers in the United States in 2008 revealed deep divisions within the Orthodox community. A third (33 percent) favored teaching creationism in the public schools, another third (35 percent) opposed the teaching of creationism, while the remainder (32 percent) could not make up their minds.[33]

Conclusion

By 2009, "the year of Darwin and Galileo," 1,631 years had passed since Basil of Caesarea, a founding father of Orthodoxy, composed his *Homilies on the Hexaemeron*, in which he spelled out the relations between the sciences and Eastern Christianity. Since Basil's day, the sciences—once the speculative occupation of a tiny minority of scholars and philosophers—have become (along with technology) a defining characteristic of civilization and the major concern of the richest countries. Has too much changed even to make meaningful generalizations about the history of science and Eastern Orthodoxy? I think not.

Present-day Orthodoxy claims to be the heir of the Greek fathers. It bases itself on their exegetical texts, which are still studied in schools and faculties of theology—not just as historical documents but also for inspiration and doctrine. This commitment, however, does not necessarily reduce the value of secular knowledge. Since antiquity, many Orthodox scholars have appreciated the sciences for providing rational explanations of the world and for offering insights into questions of faith. How can rationality and revelation be reconciled? To what extent can reason explain the mysteries of the universe? What scientific importance should be given to the texts of the fathers? For some contemporary Orthodox scholars, the Big Bang cosmogony demonstrates the accuracy of the fathers' texts, because of its similarity to the fathers' conception of the birth ex nihilo of space, time, and matter.

Since Origen, the third-century theologian, Eastern Christianity has taken an interest in science. The idea that prevailed—though not without opposition—was that the main concern of believers should be the purification of their souls in order to glorify the marvel of Creation. This Creation, however, should be compatible with the image of the world described by the philosophers. Two factors facilitated the adoption of this position: the Greek language and institutional continuity. Eastern Christianity spoke and wrote in Greek, the language of the philosophers and mathematicians, and was an integral part of the Eastern Roman Empire, which inherited the schools of antiquity. Thus, although interest in science declined sharply in the first Christian centuries, science education was never ignored, as happened to Western Christianity in the early Middle Ages. Even if in some periods schools declined or closed, overall the secular Byzantine power structure maintained an educational system that included science.

Although secular leaders kept science teaching alive, the church often played a supporting role. The great ecclesiastical debates during the Byzantine period constitute landmarks for the relations between Orthodoxy and science. Five centuries after Basil and Gregory of Nyssa reconciled Christianity with the science of the ancient Greeks and saved the prestige of secular learning, the debate over icons again posed the question of the necessity of such learning. The closing of the university during this period proved ephemeral, and Orthodox scholars increasingly saw themselves as the inheritors of ancient Greek science. These developments led to the revival of secular learning during the ninth century, the period of "Byzantine humanism." The Hesychasts of the fourteenth century, seeking the holy light of revelation through prayer and bodily exercise, once more devalued secular learning, dismissing it as ephemeral and of little help in understanding the Creation. But once again science survived, even thrived. During and just after the debate over Hesychasm, science flourished more than ever in Byzantium. In the long run, Hesychast ideas exerted a much greater influence on Slavic Orthodoxy that on Greek Orthodoxy.

After the fall of Byzantium to the Ottomans in 1453, the most important landmark in the history of Eastern Christianity, the Christian patriarch of Constantinople, came under the control of a Muslim power, while the Russian Orthodox Church started down its own independent path. The Greek church's involvement in science increased, however, because the millet system instituted by the Ottoman Empire gave the patriarch control over the education of Christians. During the five centuries of Ottoman domination, the problem of secular knowledge arose whenever a change was in view, especially during the period of "Orthodox humanism" at the beginning seventeenth century and during the Greek Enlightenment in the mid-eighteenth century. In both instances, the Orthodox Church was divided between conservatives and the partisans of reform. In the first case, the reformers supported scientific teaching in the curriculum of the schools, following the ideas of humanism, which promoted secular learning. In the second, the progressives promoted the introduction of the new European science associated with the "Scientific Revolution" in order to support the idea of national revival. In both cases, the reformers emphasized their intellectual ties to ancient Greece.

The most significant characteristic that differentiates the history of science in the Eastern Orthodox world from what happened in the Latin West (through the nineteenth century) is the East's continuing pride in its ancient Greek patrimony. Although "Hellene" was synonymous with "pagan," the Greek fathers based their Creation exegesis on their Greek education; later, Byzantine scholars (most of whom were clerics) regarded it as an honor to be "Hellene." Greek Orthodox communities of the Ottoman Empire, seeking a national identity, claimed their affiliation with the ancient Greeks. Through the centuries, this affiliation gave rise to a relatively stable relationship between Eastern Orthodoxy and science, during which the Orthodox Church accepted and taught science.

In terms of institutional support for science, the Orthodox Church also departed from practices in the West. Until the late Middle Ages, learning in Western Christianity remained largely in the hands of the monasteries, and the Roman Catholic Church often played an important role in the founding of the universities in the late eleventh and twelfth centuries. The Jesuits especially supported astronomy and mathematics in their colleges. In the East, during the Byzantine period, secular financial support for scientific learning seems to have been more important than ecclesiastical support. During the Ottoman period, Orthodox communities organized numerous schools, but these institutions were rarely financed directly by the church. Nevertheless, it was only at the beginning of the nineteenth century that this financial independence led to a relative curricular independence from church control.

Another important turning point in the history of Eastern Orthodoxy was the breakup of the Ottoman Empire at the end of the nineteenth century, which led to the constitution of independent national Orthodox churches in southeastern Europe. Though doctrinally similar, they sometimes differed in their attitudes toward science, often reflecting the policies of their national states. In contrast to some segments of Western Christianity, the Eastern Orthodox Church seemed always to be struggling against a state that was trying to control it: the Byzantine Empire and then the Ottoman Empire for the Patriarchate of Constantinople, the Russian Empire for the Moscow Patriarchate, and the national states for the independent national churches of southeastern Europe. These experiences fostered a tendency to adapt to prevailing political and ideological circumstances. The national Orthodox churches' moderate and flexible attitude with respect to a contested issue such as the theory of evolution is just one example of their attempts to maintain a balance between Orthodox dogma and state strategies.

Historiographically, there has been a retrogression of interest in Eastern Orthodoxy. The two most influential nineteenth-century polemics on science and religion, John William Draper's *History of the Conflict between Religion and Science* (1874) and Andrew Dickson White's *A History of the Warfare of Science with Theology in Christendom* (1896), paid more attention to Eastern Orthodoxy than do many recent studies. Although Draper perpetuated the story of Hypatia's brutal murder at the hands of Christians in Alexandria and chastised Cosmas Indicopleustes for advocating a flat earth, he generally treated Eastern Orthodoxy (and Islam) with respect, especially compared with his unrelenting attack on the Roman Catholic Church. White, who spent several years in the early 1890s in St. Petersburg as the United States ambassador to Russia, inexplicably overlooked Hypatia but devoted even more attention than Draper to the fallacious teachings of Cosmas (probably because this latter had an important influence in Russia). White included discussions of the church fathers' texts on the Creation, of John of Damascus's interpretation of comets in the eighth century, of the twelfth-century Greek church's views on usury, and of the seventeenth-century Russian Orthodox patriarch Nikon's interpretation of comets as divine portents. Unlike Draper, who never mentioned the Eastern Church after the Middle Ages, White referred to its attitude toward biblical interpretation in the eighteenth century and to the "Greek church" in Russia using scripture to forbid peasants from raising and eating potatoes. In his autobiography, he included a lengthy tale of the Russian church's credulity about miracles.[1]

The paucity of attention paid to Eastern Orthodoxy in histories today can be seen in a quick survey of the most general recent literature. John Hedley Brooke's influential *Science and Religion: Some Historical Perspectives* (1991) says nothing about Orthodoxy. *God and Nature: Historical Essays on the Encounter between Christianity and Science* (1986), edited by David C. Lindberg and Ronald L.

Numbers, devotes only a page or so to the Greek church fathers, as does their more recent *When Science and Christianity Meet* (2003). Gary B. Ferngren's *The History of Science and Religion in the Western Tradition: An Encyclopedia* (2000) includes a five-page overview of "Orthodoxy"—but it took four scholars to write it. There is no sign that the situation is improving. Ronald L. Numbers's iconoclastic *Galileo Goes to Jail and Other Myths about Science and Religion* (2009) says nothing about Orthodoxy except for a two-page debunking of the Hypatia myth. Despite its comprehensive title, Peter Harrison's *The Cambridge Companion to Science and Religion* (2010) completely ignores Eastern Christianity—except to apologize for doing so—while Thomas Dixon, Geoffrey Cantor, and Stephen Pumfrey, the editors of *Science and Religion: New Historical Perspectives* (2010), don't even apologize for their neglect. Perhaps the most surprising omission appears in John Hedley Brooke and Ronald L. Numbers's cutting-edge *Science and Religion around the World* (2010), which devotes chapters to Judaism, Christianity, Islam, Hinduism, Buddhism, indigenous African religions, and unbelief but mentions Eastern Orthodoxy only in passing. This hardly seems appropriate for one of the largest Christian communions in the world, second only to the Roman Catholic Church.[2]

This paucity of interest in Orthodoxy is less apparent in recent histories of Christianity. The nine volumes of the *Cambridge History of Christianity* (2006–8) devote considerable attention to Eastern Christianity. In addition to volumes 1 (*Origins to Constantine*) and 2 (*Constantine to 600*), which focus on periods when the greatest part of Christianity belonged to the East, volume 5 focuses exclusively on Eastern Christianity, and volume 3 (*Early Medieval Christianities*) includes two chapters on Eastern churches. A classic book on the subject is Bishop Kallistos Ware's *The Orthodox Church* (1963): it describes the Byzantine Church, but after the fall of Byzantium it features the Russian church, neglecting the Greek patriarchates. Other influential studies are Joan M. Hussey's *The Orthodox Church in the Byzantine Empire* (1986) and Steven Runciman's *The Great Church in Captivity: A Study of the Patriarchate of Constantinople from the Eve of the Turkish Conquest to the Greek War of Independence* (1968), the broadest overview. Sadly, none of these books discusses Orthodoxy and science in any detail.[3]

The first to write extensively about the cosmology and physics of the fathers of the church was the distinguished French physicist and historian of science Pierre Duhem (1861–1916). In his monumental ten-volume work, *Le système du monde, histoire des doctrines cosmologiques de Platon à Copernic* (1913–59), he devotes more than a hundred pages to the early church fathers, especially the three most influential Greeks—Basil, Gregory of Nyssa, and John Chrysostom—and the Latins Ambrosius and Augustine. A devout Catholic, Duhem explored the efforts of these fathers to view the Creation through the eyes of Greek natural philosophy and astronomy. Although historians of philosophy and of science have provided useful studies concerning the physics of the creation—for example, of Philo (David T. Runia, *Philo of Alexandria and the* Timaeus *of Plato*, 1986), Philopon (Richard Sorabji, ed., *Philoponus and the Rejection of Aristotelian Science*, 1987), and Cosmas Indicopleustes (Wanda Wolska-Conus, *Cosmas Indicopleustès, Topographie Chrétienne*, 1968)—almost a century after the death of Duhem, no exhaustive study on the science of the Greek fathers yet exists, in any language. John F. Callahan's "Greek Philosophy and the Cappadocian Cosmology" (1958) and my own "La cosmologie 'savante' de l'église chrétienne orientale" (2006) offer only brief introductions to the subject. In his book *Light from the East* (2003), Alexei V. Netseruk approaches the fathers' cosmology and natural philosophy mainly from a theological rather than a historical point of view.[4]

The paucity of bibliography pertains to the whole Byzantine period: historians of science have rather neglected the history of Byzantine science. And, by and large, they have perpetuated an image of an empire where theology inhibited science for more than a millennium. The earliest studies along this line were those of Paul Tannery, mainly on mathematics (*Mémoires scientifiques* IV, 1920), and of Michael Stephanides on alchemy. The founding father of the history of science, George Sarton, devoted a total of only nine pages to Byzantium in his *Introduction to the History of Science* (1927–48). René Taton's multivolume *Histoire générale des sciences* (1957) contains a small chapter of twelve pages on Byzantine science, focusing essentially on medicine. *The Cambridge Medieval History* volume on the Byzantine Empire (1967) tried for the first time to fill the gap with a chapter by Kurt Vögel on science in Byzantium. Ten years later, Herbert Hunger, in his monumental work *Die hochsprachliche profane Literatur der Byzantiner* (1978), allotted three chapters to Byzantine science (mathematics, astronomy, and astrology; natural sciences; and medicine).

Thus, we still have no overview of the history of Byzantine science. A large number of important manuscripts remain unpublished and unstudied. Astronomy has received the most attention. To date, Anne Tihon has supervised the preparation of nine volumes of Byzantine astronomical texts. She has also written article-length overviews of Byzantine astronomy. Recently we have seen growing interest in magic and astrology in Byzantium—for example, Paul Magdalino and Maria Mavroudi's edited volume *The Occult Sciences in Byzantium* (2006) and Magdalino's *L'orthodoxie astrologue: la science entre le dogme et la divination à Byzance, VIIe–XIVe siècle* (2006). Other books shedding light on the history of Orthodoxy and science include Paul Lemerle's *Le premier humanisme byzantin* (1971), a classic on the scientific revival of the ninth and tenth centuries; Costas N. Constantinides' *Higher Education in Byzantium in the 13th and early 14th Centuries* (1982), which focuses on science education and its implications for the church; John Meyendorff's studies on Hesychasm (1974); and Basil Tatakis's *La philosophie byzantine* (1949), a landmark overview of Byzantine philosophy that provides essential background for science/Orthodoxy studies.[5]

Strangely enough, we know much more about the history of science in the Orthodox communities of the Ottoman Empire than about Byzantine science. Since the 1970s, Greek historians of science have focused considerable attention on the reception of modern European science during the seventeenth and eighteenth centuries. They have inquired into how Western "Catholic" or "Protestant" science spread to the East and explored the mechanisms of this spread and the reactions to it, including how it influenced social and religious life. This historiographical trend coincided with the return to Greece of a new generation of historians of science who had studied in Europe and America; they helped to place the Greek experience within the mainstream of the history of science. Yannis Karas, author of *Exact Sciences in the Greek World, 15th–19th Century* (1991) and editor of *History and Philosophy of Science in the Greek World 17th–19th Century* (2003), was the first of the group. I presented an overview of the history of Greek science, in French, in *L'Europe des sciences* (2001). Since clerics typically taught science during the period of Ottoman domination, most of the books presenting the history of science during that period necessarily pay some attention to science/Orthodoxy relations. Vasilios Makrides has written articles and books on science and Orthodoxy in southeastern Europe, focusing on the eighteenth to twentieth century (and especially on the polemics on the heliocentric debate and on the problem of rationality).[6]

NOTES

1. John William Draper, *History of the Conflict between Religion and Science* (New York: D. Appleton, 1874), pp. 55–56 (Hypatia), 68–75, 91–95 (Eastern Christianity), 154 (Cosmas); Andrew Dickson White, *A History of the Warfare of Science with Theology in Christendom*, 2 vols. (New York: D. Appleton, 1896), 1:6, 53 (fathers), 1:93–95, 104 (Cosmas), 1:175 (John of Damascus), 1:182 (Nikon), 1:236 (Genesis in the nineteenth century), 2:236 (nineteenth century), 2:265–68 (usury), 2:285 (potatoes); 2:301

(Chrysostom), 2:311 (eighteenth century); *Autobiography of Andrew Dickson White*, 2 vols. (New York: Century, 1905), 2:67–69 (miracles).

2. John Hedley Brooke, *Science and Religion: Some Historical Perspectives* (Cambridge: Cambridge University Press, 1991); David C. Lindberg and Ronald L. Numbers, eds., *God and Nature: Historical Essays on the Encounter between Christianity and Science* (Berkeley and Los Angeles: University of California Press, 1986), pp. 23ff.; David C. Lindberg and Ronald L. Numbers, eds., *When Science and Christianity Meet* (Chicago: University of Chicago Press, 2003), pp. 11–12; Allyne L. Smith Jr., H. Tristram Engelhardt Jr., Edward W. Hughes, and John Henry, "Orthodoxy," in Gary B. Ferngren, ed., *The History of Science and Religion in the Western Tradition: An Encyclopedia* (New York: Garland, 2000), pp. 268–73; Ronald L. Numbers, ed., *Galileo Goes to Jail and Other Myths about Science and Religion* (Cambridge, Mass.: Harvard University Press, 2009), pp. 8–9; Peter Harrison, ed., *The Cambridge Companion to Science and Religion* (Cambridge: Cambridge University Press, 2010), p. 16; Thomas Dixon, Geoffrey Cantor, and Stephen Pumfrey, eds., *Science and Religion: New Historical Perspectives* (Cambridge: Cambridge University Press, 2010); John Hedley Brooke and Ronald L. Numbers, eds., *Science and Religion around the World* (New York: Oxford University Press, 2010).

3. *The Cambridge History of Christianity* (Cambridge: Cambridge University Press): vol. I, *Origins to Constantine*, ed. Margaret M. Mitchell and Frances M. Young (2006); vol. II, *Constantine to 600*, ed. Augustine Casiday and Frederick W. Norris (2007); vol. III, *Early Medieval Christianities*, ed. Thomas F. X. Noble and Julia M. H. Smith (2008), including Andrew Loth, "The Emergence of Byzantine Orthodoxy," pp. 46–64, and Igor Dofmann-Lazarev "Eastern Christianities from the Persian to the Turkish Conquest," pp. 65–84; Kallistos Ware, *The Orthodox Church* (Harmondsworth: Penguin Books, 1963; new ed., fully revised, Harmondsworth: Penguin Books, 1993); Joan M. Hussey, *The Orthodox Church in the Byzantine Empire* (Oxford: Clarendon Press, 1986; new ed., Oxford: Oxford University Press, 2010); Steven Runciman, *The Great Church in Captivity: A Study of the Patriarchate of Constantinople from the Eve of the Turkish Conquest to the Greek War of Independence* (Cambridge: Cambridge University Press, 1968).

4. Pierre Duhem, *Le système du monde*, vol. II (Paris: Hermann, 1913; 2nd ed., 1974); David T. Runia, *Philo of Alexandria and the Timaeus of Plato*, Philosophia Antiqua 44 (Leiden, 1986); Richard Sorabji, ed., *Philoponus and the Rejection of Aristotelian Science* (London: Duckworth, 1987); Wanda Wolska-Conus, *Cosmas Indicopleustès, Topographie Chrétienne*, Sources Chrétiennes 141 (Paris, 1968); John F. Callahan, "Greek Philosophy and the Cappadocian Cosmology," *Dumbarton Oaks Papers*, no. 12 (1958): 31–55; Efthymios Nicolaidis, "La cosmologie 'savante' de l'église chrétienne orientale," in Mustafa Kacar and Zeynep Durukai, eds., *Essays in Honor of Ekmeleddin Ihsanoglu*, vol. 1 (Istanbul: IRCICA, 2006), pp. 475–99; Alexei V. Nesteruk, *Light form the East* (Minneapolis: Fortress Press, 2003).

5. Paul Tannery, *Mémoires scientifiques*, vol. IV, *Sciences exactes chez les byzantins* (Toulouse: Edouart Privat, 1920); Michel Stephanides, *L'art psammurgique et la*

59. *Hex.*, H6:9, p. 98. Louis Brehier connects this idea with the Stoics (see *Hex.*, p. 372, n. 4).

60. Plato developed various theories in his books: in general, vision results from two luminous vectors, one from the eye and one from the seen object. For vision in Hipparchus, cf. Plutarch, Αρεσκοντα τοις φιλοσοφοις [De placitis philosophorum], 4, 13.

61. Here, Gregory used the word δύναμης (force) instead of ποιότης (quality).

62. *Apol.*, p. 73.

63. Even for Basil, it was more a matter of illumination than of creation of light, contrary to the corresponding title Giet gives in *Hex.*, p. 171.

64. *Hex.*, H2:8, p. 33.

65. *Hex.*, H2:8, p. 34.

66. In effect, because of the sun's apparent movement, the solar day (solar midday to solar midday) is not the same as the sidereal day (duration of the Earth's revolution and, hence, in the geocentric system, the duration of a revolution of the starry sphere).

67. *Apol.*, p. 76.

68. στερεά (solid, firm), or στερέωμα (firmament).

69. Giet notes that Basil condemned Origen-inspired interpretations (see Giet, p. 234, n. 3).

70. *Hex.*, H3:9, pp. 51–52. Basil insisted on this point to refute the idea that the separation of waters was a figure of the separation of good powers from evil spirits.

71. *Hex.*, H3:4, pp. 42–43.

72. For example, Pierre Duhem, *Le système du monde*, vol. II (Paris: Hermann, 1914; 2nd ed., 1974), pp. 488–89.

73. *Hex.*, H3:4, p. 43.	74. *Apol.*, p. 81.
75. *Apol.*, p. 84.	76. *Apol.*, p. 86
77. *Hex.*, H3:5, p. 44.	78. *Apol.*, p. 85.

79. *Hex.*, H6:2, p. 85. Giet remarks that Basil perhaps borrowed this idea from Philo (*Hex.*, p. 280, n. 1).

80. *Hex.*, H6:2, p. 85.	81. *Hex.*, H6:2, p. 86.
82. *Hex.*, H6:3, p. 86.	83. *Hex.*, Giet, p. 340, n. 1.

84. *Hex.*, H6:3, p. 87. cf. Geminus, *Introduction aux phénomènes*, IX, 8.

85. *Hex.*, H6:9, p. 98.	86. *Hex.*, H6:10, p. 99.
87. Ibid.	88. *Apol.*, p. 120.

89. *Apol.*, p. 116.

90. Απειρον των αστρων πληθος—that is to say, "infinite the number of stars," but I think that the word απειρον does not indicate infinity, properly speaking, but a very large number, for "infinity" would be in contradiction with Gregory's whole theology.

91. *Apol.*, p. 117.

92. Φυσικην ιδιοτητα in the text.

93. For the ancients, the galaxy (Milky Way) was not composed of stars; it was a luminous phenomenon of the eighth sphere.

94. *Apol.*, p. 117.
95. Ibid.
96. *Apol.*, p. 116.
97. Here we find again the term απειρη (infinity).
98. *Apol.*, p. 116.
99. *The Works of Archimedes*, trans. Sir Thomas Heath (Cambridge, 1897).
100. *Apol.*, p. 117.
101. *Hex.*, H 4:2, pp. 56–57.

Chapter 2 · Two Conceptions of the World

1. See E. Amand de Mendieta, "Les neuf Homélies de Basile de Césarée sur l'Hexaéméron," *Byzantion* 48 (1978): 345.

2. See, for example, *The Life of Andreas Salos*, ed. J.-P. Migne, Patrologia Graeca, 111, and about that text, J. Grosdidier De Matons, "Les thèmes d'édification dans la Vie d'André Salos," *Travaux et Memoires* 4 (1970): 277–328. This tradition will also be propagated by the very literal chains of Genesis; see Anne-Laurence Caudano, "Un univers sphérique ou voûté? Survivance de la cosmologie antiochienne à Byzance (XI et XII s.)," *Byzantion* 78 (2008): 66–86.

3. John Chrysostom, Ὑπόμνημα εἰς τὴν Γένεσιν [Memorandum on the Genesis], *Homilies I–XXIII*, ed. Spyridon Moustakas (Thessalonica: Patristic Editions Gregory Palamas, 1981), p. 40 (H2, 3).

4. The Greek translation of the Septuagint uses the singular *heaven* and not the plural *heavens*.

5. John Chrysostom, Ὑπόμνημα εἰς τὴν Γένεσιν, pp. 153–55 (H6, 6).

6. *Pneuma*, wind or breath.

7. John Chrysostom, Ὑπόμνημα εἰς τὴν Γένεσιν, pp. 53–55 (H3, 1). This simplistic interpretation of Basils is connected more to Antiochian than to Alexandrian thinking.

8. Ibid., pp. 83–85 (H4, 3).   9. Ibid., p. 119 (H5, 3).
10. Ibid., p. 87 (H4, 4).   11. Ibid., p. 123, 147 (H5, 4 and H6, 5).
12. Ibid., p. 151 (H6, 5).   13. Ibid., p. 145 (H6, 3).
14. Ibid., p. 85 (H4, 3).

15. On Hypatia, see, for example, Maria Dzielska, *Hypatia of Alexandria*, trans. F. Lyra (Cambridge, Mass.: Harvard University Press, 1995).

16. Wanda Wolska-Conus, *Cosmas Indicopleustès, Topographie Chrétienne*, Sources Chrétiennes 141 (Paris, 1968), vol. I, p.16.

17. Book X of his *Christian Topography* is devoted to citations of passages from many authors, especially commentators on the Bible since Philo. See Wanda Wolska-Conus, *Cosmas Indicopleustès*, vol. III, pp. 237–313.

18. Wanda Wolska-Conus, *Cosmas Indicopleustès*, vol. I, p. 274.

19. Ibid.   20. Ibid., pp. 534–36.
21. Ibid., p. 538.   22. Ibid., p. 540.

23. Ibid., p. 398. Cosmas spoke of the "inn of the sun" evoked by the ancient traveler Pytheas of Marseille.

24. Ibid., p. 552.   25. Ibid., p. 484.

26. Ibid., vol. III, p. 124.   27. Ibid., p. 140.

28. Ibid., p. 226.   29. Ibid., pp. 220, 230.

30. Ibid., p. 282. Only fragments of Epiphanius's book have come down to us.

31. Ibid., vol. I, p. 552.   32. Ibid., pp. 560–69.

33. Ibid., vol. III, p. 22.   34. Ibid., p. 20.

35. *Jean Philopon: La Création du monde*, trans. Marie-Claude Rosset and M. H. Congourdeau (Paris: Migne, 2004), p. 33.

36. Ibid., p. 32. On Philoponus's ideas about this physics, see Richard Sorabji, ed., *Philoponus and the Rejection of Aristotelian Science* (London: Duckworth, 1987).

37. Philoponus, *La Création du monde*, p. 34.

38. Ibid., p. 44.   39. Ibid., p. 35.

40. Ibid., p. 60.   41. Ibid., p. 38; cf. Plato, *Timaeus*, 38b.

42. Philoponus, *La Création du monde*, p. 41.

43. Ibid., p. 43.   44. Ibid., pp. 47–52.

45. Ibid., p. 61.   46. Ibid., p. 223.

47. Ibid., p. 158.

48. "Aristotle, in supposing a fifth corporeal essence in heaven, received from us a sufficient refutation" (ibid., p. 127). According to Philo, the stars are made of a small amount of earth and a lot of fire (ibid., p. 188).

49. Ibid., p. 128.   50. Ibid., p. 160.

51. Ibid., p. 158.   52. Ibid., p. 159.

53. Ibid., p. 125.   54. Ibid., pp. 126–27.

55. Ibid., p. 166.   56. Ibid., pp. 135–39.

57. Ibid., p. 195.   58. Ibid., p. 186.

59. Ibid., p. 130.   60. Ibid., p. 164.

61. Ibid., p. 183. Philoponus was referring here to the reddish light of the moon during eclipses.

62. Ibid., p. 203.   63. Ibid., p. 204.

64. Ibid., p. 206.   65. Ibid., pp. 50–51.

66. Ibid., p. 212.

Chapter 3 · No Icons, No Science

1. Paul Lemerle, *Le premier humanisme byzantin* (Paris: PUF, 1971), p. 66.

2. Henry Chadwick, "Philoponus the Christian Theologian," in Richard Sorabji, ed., *Philoponus and the Rejection of Aristotelian Science* (London: Duckworth 1987), pp. 41–56.

3. This is contested by some historians; see Lemerle, *Le premier humanisme byzantin*, p. 71.

4. For the university, see Al. Kazhdan, with L. E. Sherry and Chr. Angelidi, *A*

History of Byzantine Literature (650–850) (Athens: Institute for Byzantine Research / NHRF, 1999), and also Lemerle, *Le premier humanisme byzantin*, ch. 4, pp. 74–108.

5. On Stephen of Alexandria, see Anne Tihon, "L'astronomie byzantine (du Ve au XVe siècle)," *Byzantion* 51 (1981): 607–8.

6. On this *quadrivium*, see Gianna Katsiampoura "Reception, Diffusion and Functioning of Science in Middle Byzantium and the *Quadrivium* of 1008" [in Greek] (PhD diss., University of Panteion, Athens, 2004).

7. Georgii Pisidae, *Opus sex dierum, seu, mundi opificium* (Lutetia, 1584) (first edition of the Greek text in Latin translation); also in Patrologia Graeca, 92.

8. Ibid., verses 355–59.

9. Ibid., verses 50ff.

10. Ibid., verse 15. On Pisides' astronomy, see G. Bianchi, "Sulla cultura astronomica di Giorgio di Pisidia," *Aevum* 40 (1966): 35–52.

11. Maximus the Confessor, *Capita Alia*, Patrologia Graeca, 90, col. 1405; *Epistole*, Patrologia Graeca, 91, col. 412C.

12. Basil Tatakis, *La philosophie byzantine* (Paris, 1949), pp. 73ff.

13. On writing in Byzantium, see H. Hunger, *Schreiben und Lesen in Byzanz. Die byzantinische Buchcultur* (Munich: Verlag C. H. Beck, 1989).

14. Lemerle, *Le premier humanisme byzantin*, p. 89. For education during the iconoclast period, see also Ann Moffatt, "Schooling in the Iconoclast Centuries," in Anthony A. M. Bryer and Judith Herrin, eds., *Iconoclasm* (Birmingham: John Goodman & Sons, 1977), pp. 85–92.

15. Anne Tihon, "L'astronomie à Byzance à l'époque iconoclaste," in P. L. Butzer and D. Lohrmann, eds., *Science in Western and Eastern Civilization in Carolingian Times* (Basel: Birkhäuser, 1993), pp. 182–83

16. John of Damascus, "De imaginibus oratio I," Patrologia Graeca, 94, col. 1245C. The same text appears in *Writings: The Fathers of the Church*, vol. 37, trans. Frederick Chase (Washington, D.C.: Catholic University Press of America, 1979).

17. John of Damascus, Patrologia Graeca, 94, col. 884B.

18. A description is given by Aikaterini Arabatzi, "Η κοσμολογία του Ιωάννη Δαμασκηνού" [The cosmology of John of Damascus] (master's thesis, University of Athens and National Technical University of Athens, 1999).

19. See John of Damascus, "On Light, Fire, Luminaries, the Sun, the Moon, and the Stars," Patrologia Graeca, 94, cols. 885–900.

20. Basil of Caesarea, *Exegetical Homilies*, trans. Sister Agnes Clare Way (Washington, D.C.: Catholic University of America Press, 1963); Basil of Caesarea, *Homélies sur l'Hexaéméron*, ed. by Stanislas Giet, Sources Chrétiennes 26bis (Paris, 1968), p. 349.

21. Gregory of Nyssa, *Against Fate*, www.sage.edu/faculty/salomd/nyssa/fate .html. See also translation in French in *Les Pères de l'église et l'astrologie* [Church fathers and astrology], introduction by Marie-Elisabeth Allamany, ed. J.-P. Migne, Les pères dans la foi (Paris, 2003), pp. 113–45.

22. Diodoros of Tarsus, *Against Fate*, summary of the book in Photius, *Biblio-*

thèque, codex 223, ed. R. Henry, vol. IV (Paris: Belles Lettres, 1963). See *Les Pères de l'église et l'astrologie*, p. 163.

23. Tihon, "L'astronomie à Byzance à l'époque iconoclaste," pp. 181–203.

24. Franz Rosenthal, "From Arabic Books and Manuscripts," *Journal of the American Oriental Society* 83, no. 4 (Sept.–Dec. 1963): 452–57.

25. Translation by Tihon, "L'astronomie à Byzance à l'époque iconoclaste," p. 185.

26. David Pingree, "Classical and Byzantine Astrology in Sassanian Persia," *Dumbarton Oaks Papers* 43 (1989): 227–39.

27. For Byzantine chronology, see V. Grumel, *La Chronologie,* Traités d'Études byzantines I (Paris: Presses Universitaires France, 1958).

28. On astrology in Byzantium, see Paul Magdalino, *L'Orthodoxie des astrologues: La science entre le dogme et la divination à Byzance (VIIe–XIVe siècle)*, Réalités byzantines 12 (Paris: Lethielleux, 2006).

Chapter 4 · The Return of Greek Science

1. For the Antikythera mechanism, see www.antikythera-mechanism.gr/.

2. Gianna Katsiampoura, "John Grammatikos, Scientist and/or Magus?" [in Greek], in K. Skordoulis et al., eds., *Ζητήματα επιστήμης: ιστορία, φιλοσοφία και διδακτική* [Questions of science: History, philosophy and didactics] (Athens: Nissos, 2008), pp. 29–36.

3. Paul Lemerle, *Le premier humanisme byzantin* (Paris: PUF 1971).

4. Ibid., pp. 149–50.

5. Ibid., pp. 150–52.

6. On the telegraph, see Milton Anastos, *Ιστορία του Ελληνικού Έθνους* [History of the Greek nation], vol. 8 (Athens: Ekdotiki Athinon, 1979), p. 268.

7. Christine Nomikou and Gianna Katsiampoura, "The School of Magnavra. Sciences in Byzantium" [in Greek], in Skordoulis et al., *Ζητήματα επιστήμης*, pp. 37–43.

8. Georgios Kedrenos, *Σύνοψις Ιστοριών* [Summary of histories], ed. I. Bekker (Bonnae, 1838–39), vol. II, p. 326. It is a compilation of the history of mankind since the Creation to the emperor Isaac Komnenos (r. 1057–59).

9. On Photius's library, see Warren T. Treatgold, *The Nature of the Bibliotheca of Photius*, Dumbartron Oaks Studies 18 (Washington, D.C.: Dumbarton Oaks Center for Byzantine Studies, 1980), and A. Markopoulos, "Νέα στοιχεία για τη χρονολόγηση της βιβλιοθήκης του Φώτιου" [New evidence for the dating of Photius's Library], *Byzantina Symmeikta* 7 (1987): 183–91.

10. Patrologia Graeca, 105, col. 509.

11. Basil Tatakis, *La philosophie byzantine* (Paris: Presses Universitaires de France, 1949), p. 131.

12. Lemerle, *Le premier humanisme byzantin*, pp. 172–75.

13. Anne Tihon, "L'astronomie byzantine (du Ve au XVe siècle)," *Byzantion* 51 (1981): 609.

14. Lemerle, *Le premier humanisme byzantin*, pp. 169–72. We know also that Leo

possessed works by Kyrinos and Markellos, Apollonius, Theon of Alexandria, Proclus, Ptolemy Archimedes, Euclid, as well as a number of astrological texts.

15. Ibid., p. 220.

16. On Psellos, see Christian Zervos, *Un philosophe néoplatonicien du XIe siècle: Michel Psellos, sa vie, son œuvre, ses luttes philosophiques, son influence*, preface by François Picavet (Paris, 1920; New York: B. Franklin, 1973); Anitra Gadolin, *A Theory of History and Society with Special Reference to the Chronographia of Michael Psellus: 11th Century Byzantium and a Related Section on Islamic Ethics* (Amsterdam: A. M. Hakkert, 1987).

17. On Psellos's alchemy, see Gianna Katsiampoura, "Transmutation of Matter in Byzantium: The Case of Michael Psellos, the Alchemist," *Science and Education* 17 (2008): 663–68.

18. Mstislav Antonini Sangin, *Codices Rossicos (Catalogus Codicum Astrologorum Graecorum)*, XII (Brussels: H. Lamertin 1936), p. 167.

19. E. Kurtz and F. Drexl, *Michaellis Pselli: Scripta minora*, vol. I (Milan: Societa editrice Vita e Pensiero, 1936), p. 447.

20. K. Sathas, *Annuaire de l'association pour l'encouragement des etudes grecques dans la France*, 5 (Paris, [1881]), p. 58.

21. See Kurtz and Drexl, *Michaellis Pselli*, p. 441, and J. Boissonade, *Michael Psellus: De operatione daemonum* (Nuremberg, 1838), pp. 153–54.

22. On the *quadrivium* of 1008, see chapter 3.

23. Διδασκαλία παντοδαπή, in *Michael Psellus de omnifaria doctrina*, ed. L. G. Westerink, with a critical text and introduction (Utrecht: J. L. Beijers, 1948).

24. The titles of these two texts are in Greek: Ποίημα του μακαριωτάτου Ψελλού περί της κινήσεως του χρόνου, των κύκλων του ηλίου και της σελήνης, της εκλείψεως αυτών και της του Πάσχα ευρέσεως and Περί λίθων δυνάμεως.

25. L. Clucas, *The Trial of John Italos and the Crisis of Intellectual Values in Byzantium in the Eleventh Century* (Munich: Institut für Byzantinistik der Universität München 1981).

26. On the roles of church and state in education at this time, see Robert Browning, *Church, State and Learning in the Twelfth Century Byzantium* (London: Dr Williams's Trust, 1981).

27. The daughter of Alexius, Anna Komnena (1083–1153), recounted the history of her father in her *Alexiad*, an important medieval source. The text had many editions, for example *The Alexiad of Anna Comnena*, ed. and trans. E. R. A. Sewter (Harmondsworth: Penguin, 1969).

28. According to one observation of an eclipse that he mentioned, which may well have taken place in 1086 and not in 1058. Seth's dates are subject to discussion.

29. The Greek titles are: Σύνοψις φυσικής, περί χρείας των ουρανίων σωμάτων; Σύνταγμα κατά στοιχείον περί τροφών δυνάμεων; and Περί φουκάς. On astrology and Seth, see Paul Magdalino "The Byzantine Reception of Classical Astrology," in Catherine Holmes and Judith Waring, eds., *Literacy, Education and Manuscript Transmission in Byzantium and Beyond* (Leiden: Brill, 2002), pp. 33–57.

30. Edition of *On Natural Things* (or *Synopsis of Physical Problems*) in Patrolo-

gia Graeca, 122, cols. 783– 819. In the Patrologia Graeca, this work by Seth appears under the name Psellos; also edited by A. Delatte, *Anecdota Atheniensia et alia II: textes relatifs à l'histoire des sciences* (Paris: E. Droz, 1939), pp. 1–89. See also Manolis Kartsonakis, "Η σύνοψις των Φυσικών του Συμεών Σηθ" [On natural things of Symeon Seth], in George Vlahakis and Efthymios Nicolaidis, eds., *Βυζάντιο-Βενετία-Νεώτερος ελληνισμός* [Byzantium-Venice-Modern Hellenism] (Athens: NHRF, 2004), pp. 129–37.

31. Glykas's reply: *Μιχαήλ του Γλυκά, Ανταπολογητικόν προς την εγχειρισθείσαν αυτώ γραφήν του κραταιού και αγίου ημών βασιλέως κυρού Μανουήλ του Κομνηνού* [Michael Glykas's reply to the given to him letter from our saint and omnipotent king Manuel Komnenos], in F. Cumont, ed., *Catalogus Codicum astrologorum graecorum*, vol. V, part 1 (Brussels: H. Lamertin, 1904), pp. 125–40. The work of Glykas: *Ει χρη μαθηματικήν επιστήμην αποτρόπαιον ηγήσθαι παντάπασιν*, ibid., pp. 140ff. See Fl. Evaggelatou-Notara, "Οποίον εστί μέρος της αστρολογίας κακιζόμενόν τε και αποτρόπαιον (Αστρολογία-Αστρονομία και οι σχετικές αντιλήψεις κατά τον ΙΒ΄αιώνα)" [What part of astrology is nasty and horrible (Astrology-astronomy and related concepts during the 12th c.)], in Nikos Oikonomides, ed., *Το Βυζάντιο κατά τον 120 αιώνα. Κανονικό Δίκαιο, κράτος και κοινωνία* [Byzantium during the 12th c.: Canonical law, state and society] (Athens: Etaireia Vyzantinon kai Metavyzantinon Meleton, 1991), pp. 447–463.

32. The titles of Kamateros's poems are *Περί του ζωδιακού κύκλου και των άλλων απάντων των εν ουρανώ* [On the zodiacal circle and on all the stars in the sky], and *Εισαγωγή αστρονομίας* [Introduction to astronomy]. On Kamateros's treatise of the astrolabe, see Anne Tihon, "Traités byzantins sur l'astrolabe," *Physis* 32 (1995): 323–57. For discussion of the validity of astrology in Byzantium, see Paul Magdalino, *L'orthodoxie des astrologues: La science entre le dogme et la divination à Byzance (VIIe–XIVe siècle)*, Réalités byzantines 12 (Paris: Lethielleux, 2006).

33. See P. Gautier, ed., *Michel Italikos, Lettres et discours* (Michael Italikos, letters and discourses), Archives de L'Orient Chrétien 14 (Paris: Institut Français d'Études Byzantines, 1972), p. 95.

34. Timothy S. Miller, *The Birth of the Hospital in the Byzantine Empire* (Baltimore: Johns Hopkins University Press), 1985.

Chapter 5 · Struggle for Heritage

1. On the fourth crusade, see for example, K. M. Setton, *A History of the Crusades*, II: *The Later Crusades* (Madison: University of Wisconsin Press, 1969).

2. N. Costas Constantinides, *Higher Education in Byzantium in the 13th and Early 14th Centuries* (Nicosia: Cyprus Research Centre, 1982), pp. 17–19.

3. Ibid., p. 21.

4. Ibid., p. 10.

5. M. Cacouros and M.-H. Congourdeau, eds., *Philosophie et sciences à Byzance de 1204 à 1453* (Leuven: Uitgeverij Peeters en Departement Oosterse Studies, 2006), p. 2.

6. For a study of the cosmology of the *Epitome of Physics*, see Ioanna Bouzoudi, "Nicephorus Blemmydes and His Treatise of Physics (1260): A Study of His Theory of Heaven and Its Sources" (master's thesis, Université de Lille III, October 2000).

7. Nicephorus Blemmydes, *Επιτομή φυσικής* [Epitome of physics] (Leipzig, 1784), ch. 25, "On Ether and Stars," pp. 128–40. See also p. 118.

8. Ibid., p. 119.

9. Ibid., p. 47.

10. Ibid., pp. 48–49.

11. Aristotle, *De caelo*, 294a 15–18.

12. Blemmydes, *Επιτομή φυσικής*, p. 49. For the notion of place of Syrianus, Damascius, and Simplicius, see Samuel Sambursky, *The Physical World of Late Antiquity* (London: Routledge-Kegan, 1962), pp. 2–6. For Blemmydes' discussion on this theme, see Bouzoudi, "Nicephorus Blemmydes," pp. 34–37.

13. Blemmydes, *Επιτομή φυσικής*, p. 121.

14. Ibid., pp. 127–28.

15. Ibid., p. 124.

16. Ibid., p. 125–26.

17. Philoponus, *Contra Proclum*, III, 4.

18. Blemmydes, *Επιτομή φυσικής*, p. 126. For omnipotent, Blemmydes used the word απειροδύναμος (of infinite force).

Chapter 6 · Political Debates Become Scientific

1. Sophia Mergiali, *L'enseignement et les lettrés pendant l'époque des Paléologues* (Athens: Etaireia ton filon tou laou, 1996), p. 16.

2. Costas N. Constantinides, *Higher Education in Byzantium in the 13th and Early 14th Centuries* (Nicosia: Cyprus Research Centre, 1982), p. 33.

3. On the life, work, and bibliography of Pachymeres, see Stylianos Lampakis, *Γεώργιος Παχυμέρης, Πρωτέκδικος και Δικαιοφύλαξ* [Georgios Pachymeris, Protekdikos and Dikaiophylax] (Athens: NHRF, 2004). Edition of Pachymeres *quadrivium*: George Pachymère, *Traité des quatre leçons, arithmétique, musique, géométrie et astronomie*, ed. P. Tannery and E. Stéphanou, Studi e Testi 94 (Rome: Vatican, 1940).

4. Gianna Katsiampoura, "Comparing Two Byzantine *Quadrivia*: The *Quadrivium* of 1008 and G. Pachymeres Syntagma. Resemblances and Differences," in *Atti del Convegno "Libri di Scuola e Pratiche Didattiche dall'Antichita al rinancimento"* (Cassino: Eitioni Universita degli Sudi di Cassino, 2010), pp. 410–25.

5. Pachymeres, *Traité des quatre leçons*, p. 370.

6. Weak in astronomy, he believed that the value of the precession is the difference between the duration of the real annual revolution of the sun (the sidereal year, the time the sun takes to occupy the same position in relation to the stars) and the 365 days plus a quarter that would be the length of the sidereal year without taking precession into account. Nevertheless, he gave the Ptolemaic value of one degree for every hundred years. Pachymeres, *Traité des quatre leçons*, p. 364.

7. Paul Tannery, *Mémoires scientifiques*, XIII (Correspondance) (Toulouse: Ed-

ouart Privat, 1934), p. 344. The sources of Pachymeres' *quadrivium* include Diophantus, Nicomachus of Gerasse, Euclid, Ptolemy, Aratos, Archimedes, Aristotle, Cleomedes, Eratosthenes, and Theon of Alexandria.

8. For Planoudes' life, see Constantinides, *Higher Education*, pp. 43ff.

9. Maxime Planude, *Grand calcul selon les Indiens*, ed. André Allard, Travaux de la Faculté de philosophie et lettres de l'Université catholique de Louvain, 27 (1981). Greek title of the text: Ψηφοφορία κατ᾽ Ινδούς.

10. *Principles of the Great Indian Calculation* (Αρχή της μεγάλης και Ινδικής Ψηφοφορίας), Bibliothèque Nationale, Paris, ms BN, *suppl. grec* 387. Planoudes used this treatise to write his own.

11. Theodore Metochites, "Στοιχείωσις επί τη αστρονομική επιστήμη" remains unpublished except for a few fragments. A study by B. Byden, "Theodore Metochites Stoicheiosis astronomike and the Study of Philosophy in Early Palaiologan Byzantium" (PhD diss., Göteborg University, 2001). Available on the Web: http://swepub .kb.se/bib/swepub:oai:services.scigloo.org:34134?tab2=abs&language=en.

12. Preface of the *Elements*, ed. K. Sathas, Μεσαιωνικη Βιβλιοθήκη [Bibliothèque médiévale], vol. 1 (Athens, 1872), pp. πδ΄-ριη΄. Metochites does not seem to have calculated precise eclipses, despite the fact that he said he had. He did give only the means for this calculation in his *Elements*.

13. Ihor Sevčenko, *Études sur la polémique entre Théodore Métochite et Nicéphore Choumnos* (Brussels: Editions de Byzantion, 1962), pp. 13-16.

14. The texts (public correspondence between Choumnos and Metochites and pamphlets) as well as the chronicle of this polemic and the lives of Choumnos and Metochites are presented by Ihor Senčenko, ibid.

15. Metochites, "Στοιχείωσις επί τη αστρονομική επιστήμη," ch. 3.

16. Anne Tihon, "L'astronomie byzantine (du Ve au XVe siècle)," *Byzantion* 51 (1981): 614.

17. Sevčenko, *Études sur la polémique*, p. 262.

18. Nikephoros Gregoras, *Byzantina historia*, ed. L. Schopen, Corpus scriptorium histariae byzantinae VIII (Bonn, 1829).

19. Gianna Katsiampoura, "Nikephoros Gregoras versus Barlaam of Calabria: A Debate over the Prediction of Eeclipses in Constantinople in the 14th Century" [in Greek], *Neusis* 13 (2004): 138-48.

20. J. Mogenet, A. Tihon, R. Royez, and A. Berg, *Nicéphore Grégoras, calcul de l'éclipse de soleil du 16 juillet 1330* [Calculation of the solar eclipse of 16 July 1330], Corpus des astronomes byzantins I (Amsterdam, 1983), p. 13.

21. Edition of the 1330 eclipse calculations with comments, ibid.

22. The titles of pamphlets are Αντιλογία, Φιλομαθής and Φλωρέντιος. See Katsiampoura, "Nikephoros Gregoras versus Barlaam of Calabria," pp. 147-48.

Chapter 7 · True Knowledge and Ephemeral Knowledge

1. John Meyendorff, "Les débuts de la controverse hésychaste," *Byzantion* 23 (1953): 88.

2. Jean-Yves Leloup, *Écrits sur l'hésychasme* (Paris: Albin Michel 1999), p. 24.

3. Compare this with Saint Gregory of Nyssa's theory that the created light is the element fire, as distinct from the divine or uncreated light (*aktiston fōs*) from outside Creation (see chapter 1).

4. Meyendorff, "Les débuts de la controverse hésychaste," pp. 100–101.

5. The source at issue is the *Mystical Theology* of Pseudo-Dionysius.

6. Philotheos Kokkinos, *Λόγος εις τον εν αγίοις πατέρα ημών Γρηγόριον αρχιεπίσκοπον Θεσσαλονίκης* [Discourse on our Holy Father Gregory, archbishop of Thessalonica], Patrologia Graeca, 51, col. 584, translation by John Meyendorff, "Les débuts de la controverse hésychaste," p. 105.

7. John Meyendorff, "Un mauvais théologien de l'unité au XIVe siècle: Barlaam le Calabrais," in *L'Église et les Églises, 1054–1954: Etudes et travaux offerts a Dom Lambert Baudouin,* II (Chévetogne: Éd. De Chévetogne, 1954), pp. 57–58.

8. Nicolaos Katsiavrias, "Η κοσμοαντίληψη του Αγίου Γρηγορίου του Παλαμά (1296–1359)" [The perception of the world of Saint Gregory Palamas, 1296–1359] (PhD diss., University of Athens, 2001), p. 42.

9. Gregory Palamas, *Letter to Philosophers John and Theodore,* in *Complete Works of Gregory Palamas,* vol. 8, ed. P. K. Christou (Thessalonica: Patristic editions Gregory Palamas, 1994), par. 29. For Palamas's views on science, see also Gregory Palamas, "Science Does Not Save," in *The Triads,* ed. John Meyendorff, trans. Nicholas Gendle (New York: Paulist Press, 1983).

10. Katsiavrias, "Η κοσμοαντίληψη," pp. 57–58.

11. Ibid., p. 66.

12. Gregory Palamas, *The One Hundred and Fifty Chapters: A Critical Edition,* ed. Robert E. Sinkewicz (Toronto: Pontifical Institute of Medieval Studies, 1988), par. 43.

13. Gregory Palamas, *Λόγος αντιρρητικός προς Ακίνδυνον* [Contra Akindynos], in *Complete Works of Gregory Palamas,* vol. 6, ed. P. K. Christou, critical text by Leonidas C. Contos (Thessalonica: Patristic editions Gregory Palamas, 1987), theses ΣΤ΄, 11.

14. Ibid., ΣΤ΄, 27.

15. Philotheos Kokkinos, *Λόγος,* col. 560.

16. Gregory Palamas, *Λόγος αντιρρητικός προς Ακίνδυνον,* Ζ΄ 24 (see Katsiavrias, "Η κοσμοαντίληψη," p. 216).

17. Ibid., Ζ΄, 9, 25.

18. Ibid., Ζ΄, 26.

19. Gregory Palamas, *One Hundred and Fifty Chapters,* par. 26.

20. Katsiavrias, "Η κοσμοαντίληψη," pp. 221–22.

21. Gregory Palamas, *One Hundred and Fifty Chapters,* par. 5 and 6.

22. Ibid., ch. 3.

23. Ibid., 13.

24. Ibid., 16.

25. Ibid., 20.

26. Ibid., 25.

27. Ibid., 81.

28. See, for example, John Meyendorff, *St. Gregory Palamas and Orthodox Spirituality*, trans. Adele Fiske (New York: St. Vladimir's Seminary Press, 1974), pp. 143ff.

Chapter 8 · Ancients versus Moderns

1. Anne Tihon, "L'astronomie byzantine (du Ve au XVe siècle)," *Byzantion* 51 (1981): 614.

2. It should be noted that, because of various errors by the Byzantines, tables from the Arab tradition were as prone to error as the Ptolemaic ones.

3. Anne Tihon, "Sur l'identité de l'astronome Alim," *Archives Internationales d'Histoire des Sciences* 39 (1989): 3–21; R. Mercier, "The Parameters of the Zīj of Ibn al-'Alam," *Archives Internationales d'Histoire des Sciences* 39 (1989): 22–50.

4. Εἰς τὸ ποιειν κανόνιον κατά τον Ἀλίμ, in *Seldenianus* 16, f. 140 r–v, and *Neapolitanus* II C 33, f. 430 r–v. *Seldenianus* dates from the first half of the fifteenth century and the *Neapolitanus* was copied by John Xerocalitos in 1495.

5. These tables are Κατά τὴν περίληψιν τοῦ κανονίου Ἀλίμ, in *Seldenianus* 16, ff. 114–15 v, and *Neapolitanus* II C 33, ff. 402 v–403 v.

6. Annotated edition by A. Jones, *An Eleventh-Century Manual of Arabo-Byzantine Astronomy*, Corpus des Astronomes Byzantins III (Amsterdam: Brill, 1987).

7. The most remarkable Arabism is found in the chapter "On the solar eclipse" (which is almost a translation of ibn al-Muthannā). For sinus, the author uses the word περσίκιον (coin purse), but he adds ἤτοι ευθειαν ορθότητα (meaning sinus).

8. For *Synopsis*, see A. Delatte, *Anecdota Atheniensia et alia II: textes relatifs à l'histoire des sciences* (Paris: E. Droz, 1939), pp. 46 and 124. See also the list of stars titled "Τοῦ Σήθ εκείνου" (from the late Seth), ed. and annotated by D. Pingree, "The Indian and Pseudo-Indian Passages in Greek and Latin Astronomical and Astrological Texts," *Viator* 7 (1976): 177, 192.

9. The manuscript is presented by Anne Tihon in "Tables islamiques à Byzance," *Byzantion* 60 (1990): 405–13.

10. Anne Tihon, "Traités byzantins sur l'astrolabe," *Physis* 32 (1995): 331–32.

11. On the hypothesis proposed by Raymond Mercier, see Anne Tihon, "Tables islamiques à Byzance," p. 413.

12. The title in Greek: Ἑλληνιστὶ βιβλίον αστρονομικον τοῦ Ἀπομάσαρ. For astrology, properly speaking, the book by Auguste Bouché-Leclercq, *L'astrologie grecque* [Greek Astrology] (Paris, 1899), still remains a precious source of information. See also Paul Magdalino and Maria Mavroudi, eds., *The Occult Sciences in Byzantium* (Geneva: La pomme d'or, 2006).

13. For the annotated inventory of Byzantine astronomical tables of Persian origin, see Anne Tihon, "Tables islamiques à Byzance," and Tihon, "Les tables astronomiques persanes à Constantinople dans la première moitié du XIVe siècle," *Byzantion* 60 (1990): 401–25. The texts attributed to Chioniades are found in the manuscripts *Vat gr.* 185, 191, 211, 1058 and *Laur. Gr.* 28/17. These texts are the *Zīj al-'Alā'ī* by al-Fahhād (c. 1176) based on the teaching of Shams Bukhārī, the *Zīj al-Sanjarī* by

al-Khāzinī (c. 1135), unidentified tables starting in 1093, various shorter texts and figures (including the famous figures of *Vat. gr.* 211, fols. 115–21) according to the *Tadhkira* by Nasīr al-Dīn al-Tūsī (which probably inspired Copernicus for the construction of his planetary system), the tables of the *Zīj-i Īlkhānī* by al-Tūsī, and the treatise on the astrolabe. The first has been edited by D. Pingree, *The Astronomical Works of Gregory Chioniades*, vol. 1, *The Zīj al-Alā'ī*, part 1: *Text, Translation, Commentary*, part 2: *Tables*, Corpus des Astronomes Byzantins II, 2 vols. (Amsterdam, 1985–86). The text that inspired Copernicus is reproduced in E. A. Paschos and P. Sotiroudis, *The Schemata of the Stars* (Singapore: World Scientific Publishing, 1998).

14. Partially edited by A. Delatte, *Anecdota Atheniensia et alia II: textes grecs relatifs à l'histoire des sciences* (Paris: E. Droz, 1939), pp. 263ff. For a presentation, see Tihon, "Traités byzantins sur l'astrolabe," 333–35. The Persian original is unknown.

15. For an edition of the tables and translation of the text, see R. Mercier, *An Almanac for Trebizond for the Year 1336*, Corpus des Astronomes Byzantins VII (Louvain-la-Neuve, 1994). Raymond Mercier and Anne Tihon advance the hypothesis that it was the same Manuel who possessed the Chioniades corpus who wrote this *Almanac*.

16. R. Mercier, "The Greek *Persian Syntaxis* and the *Zīj-i Īlkhānī*," *Archives Internationales d'Histoire des Sciences* 35 (1985): 436–38.

17. This treatise, very important for the history of Byzantine astronomy, remains unpublished.

18. For a translation, see Anne Tihon, "Un traité astronomique chypriote du XIVe siècle," *Janus* 64 (1977): 279.

19. For a study of these Cypriot tables, see ibid., 64 (1977): 279–308; 66 (1979): 49–81; 68 (1981): 65–127. The tables themselves remain unpublished. See also P. Tavardon, "Recherche sur l'astronomie byzantine; un aspect de la première renaissance des Paléologues" (PhD diss., Toulouse, 1987), pp. 535ff.

20. Anne Tihon, "L'astronomie byzantine à l'aube de la renaissance," *Byzantion* 66 (1996): 251–52.

21. Ibid., p. 246.

22. The first two books of the *Tribiblos* were edited and annotated by Régine Leurquin, *Théodore Méliténiote, Tribiblos astronomique, livre I*, Corpus des astronomes byzantins IV (Amsterdam, 1990), *livre II*, Corpus des astronomes byzantins V–VI (Amsterdam, 1993). The third part has not yet been published.

23. Manuscript *Laur. Gr.* 28/14. See D. Pingree "The Astrological School of John Abramius," *Dumbarton Oaks Papers* 25 (1971): 191–215. Pingree suggests that the author of this list is John Abramios.

24. *Vat. gr.* 1059, fols. 251–53, *Urbinas gr.* 80, fols. 105v–106.

25. For a reference work on these translations, see C. H. Haskins, *The Renaissance of the Twelfth Century* (Cambridge, Mass.: Harvard University Press, 1927).

26. Edition and commentary by D. Pingree, "The Byzantine Version of Toledan Tables: The Work of George Lapithes?" *Dumbarton Oaks Papers* 30 (1976): 87–132.

27. Tihon, "Traités byzantins sur l'astrolabe," pp. 341–43.

28. Tihon, "L'astronomie byzantine à l'aube de la renaissance," p. 264.

29. See P. Solon, "The *Six Wings* of Immanuel Bonfils and Michael Chrysokokkes," *Centaurus* 15 (1970): 1–20.

30. See Tihon, "L'astronomie byzantine à l'aube de la renaissance," p. 254.

31. See ibid., p. 265.

Chapter 9 · The Fall of the Empire and the Exodus to Italy

1. On Illuminationism, see John Walbridge, *The Wisdom of the Mystic East: Suhrawardi and Platonic Orientalism* (Albany: State University of New York Press, 2001).

2. Anne Tihon and Raymond Mercier, *Georges Gémiste Pléthon Manuel d'Astronomie*, Corpus des astronomes byzantins IX (Louvain-la-Neuve, 1998), p. 6.

3. Ibid.

4. Ibid., pp. 269–70.

5. Theodore of Gaza, *De Mensibus*, Patrologia Graeca, 19, col. 1168 B; passage translated in Tihon and Mercier, *Georges Gémiste Pléthon*, p. 178.

6. Scholarios, *Oeuvres complètes*, ed. L. Petit, A. Sideridès, and M. Jugie, vol. IV (Paris: Maison de la bonne presse, 1931), p. 162.

7. On the Patriarchal School after the Byzantine period, see Tasos Gritsopoulos, *Πατριαρχική Μεγάλη του Γένους Σχολή* [The Patriarchal School] (Athens: Vivliothiki tis en Athinais Filekpaideutikis Etaireias, 1966).

8. Henri Omont (1857–1940, French archivist and librarian and a specialist on Greek manuscripts) estimated that in 1687, 200 Greek manuscripts were kept in Topkapi. We have evidence of the sale of manuscripts to Westerners before that date (H. Omont, *Inventaire sommaire des manuscrits grecs de la Bibliothèque Nationale* [Paris: Ernest Leroux, 1898], pp. IX, XVII). Fewer than 80 remained in 1920, when the catalog of non-Muslim manuscripts was created, of which 52 (42 in Greek) date from the time of Mehmed II; among the latter, two-thirds were books on geography, a subject of high interest for the Conqueror.

9. Copernicus would be inspired by manuscripts of Proclus or of the Persian astronomical school of Maragha translated into Greek. See, for example, N. M. Swerdlow and O. Neugebauer, *Mathematical Astronomy in Copernicus' De Revolutionibus*, 2 vols. (New York: Springer-Verlag, 1984), 1:47–48 and 2:567–68; E. A. Paschos and P. Sotiroudis, *The Schemata of the Stars* (Singapore: World Scientific Publishing, 1998); I. N. Veselovsky, "Copernicus and Nasīr al-Dīn al-Tūsī," *Journal of the History of Astronomy* 4 (1973): 128–30.

10. On Manutius and the Greek community in Venice and Greek schools in Italy, see Ambroise Firmin-Didot, *Alde Manuce et l'hellénisme à Venise* (Paris, 1875; repr. Brussels: Culture et Civilisation, 1966).

11. Kostas Petsios, *Η περί φύσεως συζήτηση στη νεοελληνική σκέψη* [The discussion on nature in modern Greek thought] (Jannina, 2003), p. 65. This book constitutes a general introduction to the natural sciences in modern Greek thought in the sixteenth and seventeenth centuries. See also Yannis Karas, *Οι θετικές επιστήμες στον ελληνικό χώρο, 15ος–19ος αιώνας* [Exact sciences in the Greek world, 15th–19th c.] (Athens: Daidalos / Zacharopoulos, 1991).

12. Manuscript of 1325 in the Monastery of Saint Laura, Mount Athos. For a complete bibliography of the scientific texts written in the Greek language during the period 1453–1821, see Yannis Karas, *Οι επιστήμες στην Τουρκοκρατία. Χειρόγραφα και έντυπα* [Sciences under Ottoman domination: Printed books and manuscripts], vol. 1, *Mathematics* (1992); vol. 2, *Natural Sciences* (1993); vol. 3, *Life Sciences* (1993) (Athens: Estia).

13. Michael Chrysokokkes translated in 1435 the astronomical tables of Immanuel ben Jacob Bonfils, Jew of Tarascon, *Kanfe nesharim* [Six wings], written around 1365 (see chapter 8). The *Physiologia* was often reprinted in Venice (first edition probably in 1603).

Chapter 10 · A Rebel Patriarch

1. On Cyril Loucaris, see Gunnar Hering, *Das ökumenische Patriarchat und europäische Politik, 1620–1638* (Wiesbaden: Franz Steiner, 1968). On education at the Patriarchal School and Loucaris, see Dimitris Dialetis, Kostas Gavroglu, and Manolis Patiniotis, "The Sciences in the Greek Speaking Regions during the 17th and 18th Centuries," in Kostas Gavroglu, ed., *The Sciences at the Periphery of Europe during the 18th Century*, New Studies in the History and Philosophy of Science and Technology, Archimedes, vol. 2 (Dordrecht: Kluwer Academic Publishers, 1999), pp. 41–71.

2. On Korydaleus's work and biography, see Cl. Tsourcas, *Les débuts de l'enseignement philosophique et de la libre pensée dans les Balkans: La vie et l'œuvre de Théophile Corydalée (1570–1646)* (Bucharest, 1948; 2nd ed., Thessalonica: Institute for Balkan Studies, 1967). See also Constantin Noica, "La signification historique de l'œuvre de Théophile Corydalée," *Revue des Etudes de Sud-Est Européennes* 2 (1973): 285–306; and Kostas Petsios, *Η περί φύσεως συζήτηση στη νεοελληνική σκέψη* [The Discussion about Nature in Modern Greek Thought] (Jannina, 2003).

3. Petsios, *Η περί φύσεως*, p. 178.

4. Theophilus Korydaleus, *Είσοδος φυσικής ακροάσεως κατ'Αριστοτέλην* (Venice, 1779), p. 5. On the preface of this book, see Nikos Psimmenos, "Η 'εκ παλαιών και νεωτέρων ερανισθείσα' Προθεωρία της 'Εισόδου Φυσικής Ακροάσεως' του Θεοφίλου Κορυδαλέως" [The prologue of the *Introduction to Aristotle's Physics* by Th. Korydaleus], in *Πρακτικά Πανελλήνιου Επιστημονικού Συνεδρίου "Το αίτημα της διεπιστημονικής έρευνας. Οι επιστήμες στον ελληνικό χώρο"* [The question of interdisciplinary research. Sciences in the Greek world] (Athens: NHRF, 1999).

5. Korydaleus, *Είσοδος φυσικής*, p. 25.

6. Theophilus Korydaleus, *Είσοδος φυσικής ακροάσεως κατ'Αριστοτέλην* [Introduction to Aristotle's *Physics*] (Venice 1779); *Γενέσεως και φθοράς κατ'Αριστοτέλην* [On genesis and decay according to Aristotle] (Venice, 1780).

7. Alexandro Mavrocordato, *Pneumaticum instrumentum circulandi sanguinis sive de motum et usu pulmonum* (Bononiae, 1664); the dissertation was reprinted in 1665, 1682, and 1870 (the last for historical reasons).

8. Ms 2846 (272), Monastery Docheiarios, Mount Athos.

9. Ms 267, Patriarchal Library of Jerusalem.

10. Petsios, *Η περί φύσεως*, p. 202.

11. Ibid., p. 203.

12. The title of the manuscript is "Astronomical book presenting and explaining the discoveries of the Ancients and the Moderns from Adam to Ptolemy and Copernicus . . ." Nine manuscripts are known; the most complete is the Collection of the Metochion of the Patriarchate of Jerusalem in Constantinople, ms 420.

Chapter 11 · Toward Russia

1. Yakov Rabkin and Sumitra Rajapopalan, "Les sciences en Russie: entre ciel et terre," in Michael Blay and Efthymios Nicolaidis, eds., *L'Europe des sciences* (Paris: Seuil, 2001), p. 222.

2. For early science and pseudoscience in the Rus and Orthodox Slavs, see, for example, Robert Romanchuk, *Byzantine Hermeneutics and Pedagogy in the Russian North: Monks and Masters at the Kirillo-Belozerskii Monastery, 1397–1501* (Toronto: University of Toronto Press, 2007); Anne-Laurence Caudano, *"Let There Be Lights in the Firmament of Heaven": Cosmological Depictions in Early Rus*, Palaeoslavica, 14, Supplementum 2 (Cambridge, Mass., 2006); William F. Ryan, *Russian Magic at the British Library: Books, Manuscripts, Scholars, Travelers* (London: British Library, 2006); Robert Mathiesen. "Magic in Slavia Orthodoxa: The Written Tradition," in Henry Maguire, ed., *Byzantine Magic* (Washington, D.C.: Dumbarton Oaks Research Library and Collection, 1995), pp. 155–78; Ihor Sevčenko, "Remarks on the Diffusion of Byzantine Scientific and Pseudo-scientific Literature among the Orthodox Slavs," *Slavonic and East European Review* 59 (1981): 321–45. For a more general approach, see Dimitri Obolensky, *The Byzantine Commonwealth: Eastern Europe, 500–1453* (New York: Praeger, 1971).

3. On Ligarides, see Harry T. Hionides, *Paisius Ligarides* (New York: Twayne, 1972).

4. See Chr. Papadopoulos, *Οι πατριάρχαι Ιεροσολύμων ως πνευματικοί χειραγωγοί της Ρωσσίας κατά τον ΙΖ΄ αιώνα* [The Patriarchs of Jerusalem: Spiritual Guides of Russia in the 17th Century] (Jerusalem, 1907).

5. See ibid. and W. Palmer, *The Patriarch and the Tsar*, vol. III (London: Trubner, 1873).

6. Hionides, *Paisius Ligarides*, p. 84.

7. Iberia was an administrative unit of Byzantium. It comprised the lands of Georgia and Armenia.

8. See B. Fonkich, *Grechko-russkie kulturnye sviazi v XV–XVII vv. Grecheshkie rukopisi v Rosii* [Greco-Russian cultural relations in XVth–XVIIth c. Greek manuscripts in Russia] (Moscow, 1972).

9. On the Academy and the Leichoudis brothers, see S. Smirnov, *Istoria Moskovskoj Slavjano-greko-latinskoj Akademmii* [History of the Slavo-Greco-Latin Academy] (Moscow, 1855); D. Yalamas, "The Students of the Leichoudis Brothers at the Slavo-Greco-Latin Academy of Moscow," *Cyrillomethodianum* 15–16 (1991–92): 113–

44; Nikolaos A. Chrissidis, "Creating the New Educated Elite: Learning and Faith in Moscow's Slavo-Greco-Latin Academy, 1685–1694" (PhD diss., Yale University, 2000).

10. Nikolaos A. Chrissidis, "A Jesuit-Aristotle in Seventeenth-Century Russia: Cosmology and the Planetary System in the Slavo-Greco-Latin Academy," in Jarmo Kotilaine and Marshall Poe, eds., *Modernizing Muscovy: Reform and Social Change in Seventeenth-Century Russia* (London: Routledge Curzon, 2004), pp. 391–416.

11. Edited by Apostolos Tsakoumis in E. Nicolaidis, ed., *Οι μαθηματικές επιστήμες στην Τουρκοκρατία* [Mathématical sciences during the Ottoman domination] (Athens: INR/NHRF, 1992), pp. 147–222.

12. For the biography of Spathar, see E. Picot, "Nicolas Spathar Milescu," in E. Legrand, ed., *Bibliographie hellénique* (Paris, 1896; repr., Brussels: Culture and Civilisation, 1963), 4:62–104; P. Panaitescu, "Nicolas Spathar Milescu," *Mélanges de l'École Roumaine en France* 1 (1925): 33–180.

13. Beate Hill-Paulus, *Nikolaj Gavrilovic Spatharij (1636–1708) und seine Gesandtschaft nach China* (Hamburg-Tokyo: Gessellschaft für Natur und Völkerkunde Ostasiens, 1978).

14. On the manuscript and its history in the Orthodox world, see Noël Golvers and Efthymios Nicolaidis, *Ferdinand Verbiest and Jesuit Science in China: An Annotated Edition and Translation of the Constantinople Manuscript (1676)* (Athens: NHRF–F. Verbiest Institute, 2009).

15. The official report by Spathar was published by Arseniev, *Puteshestvye cherez Sibir ot Tobols'ka do Nerchinski i granits Kitaya . . .* (St. Petersburg, 1882). It was translated into English by J. F. Baddeley in *Russia, Mongolia, China, Being Some Record of the Relations between Them from the Beginning of the XVIIth Century to the Death of the Tsar Alexei Michailovich, A.D. 1602–1676 . . .* (London: Macmillan, 1919), 2:286–423. The complete title of *Description of Asia* is *Opisanie pervyja chasti vselennyja imenyemoj Azii, v nej zhe sostoit Kitajskoe gosudarstvo proshchimi ego gorody i provintsii.* In fact, this book is a translation of that of M. Martini, *Novus Atlas Sinensis* (1655).

16. *Kitaia Douleuousa* [China under the yoke]. Chrysanthos is referring to the Manchurian domination.

17. Golvers and Nicolaidis, *Ferdinand Verbiest*, p. 62.

18. That year, a magnificent observatory modeled on the Islamic observatories (Maragha, Samarkand) was constructed in Constantinople, financed by the sultan. Unfortunate events such as plague and the death of many dignitaries were interpreted by the muftis as resulting from its creation, and they persuaded the sultan to order its demolition, a decision in which rivalries among high dignitaries at the court seemed to have played a role. The observatory was completely demolished on 22 January 1580, some six months after its construction.

19. For the spread of the European science in Russia, see Y. Rabkin and S. Rajapopalan, "Les sciences en Russie."

Chapter 12 · Who Were the Heirs of the Hellenes?

1. Emile Legrand, *Epistolaire grec ou recueil des lettres adressées par la plupart à Chrysante Notaras* (Paris, 1888), pp. 11–12. Chrysanthos himself mildly criticized Komninos by writing that in his teaching were found elements "according to the dogma of the Latins rather than according to the truth." See Papadopoulos Kerameus, Ιεροσολυμιτική Βιβλιοθήκη [Library of Jerusalem]) (Athens, 1899; repr., Brussels, 1963), 4:321.

2. Chrysanthos's study notes contain nothing on the new scientific ideas of these days (see codex 429, Fonds *Methochion* in Constantinople of the Patriarchate of Jerusalem).

3. Chrysanthos Notaras, Εισαγωγή εις τα γεωγραφικά και σφαιρικά [Introduction to geography and to the sphere] (Paris, 1716), p. 92.

4. G. L. Arvanitakis, "Notes astronomiques," *Le Messager d'Athènes*, no. 5217 (February 1939).

5. On the introduction of "new science" to the Orthodox world, see Efthymios Nicolaidis, "The Spread of New Science to Southeastern Europe: During or before the Greek Enlightenment?" in Celina Lertora Mendoza, Efthymios Nicolaidis, and Jan Vandersmissen, eds., *The Spread of the Scientific Revolution in the European Periphery, Latin America and East Asia*, Proceedings of the XXth International Congress of History of Science, vol. V (Turnhout, Belgium: Brepols, 1999), pp. 33–45. Neōterai epistēmai is often translated as "modern sciences." I have instead chosen "new" to emphasize the concept of novelties also expressed by this term.

6. Chrysanthos Notaras, Εισαγωγή. According to the title page, the book was printed in Paris, but it is very probable (from the preface) that it was printed in Venice and that "Paris" was falsely added by the publisher for fiscal reasons.

7. On the life of Chrysanthos, see Pinelopi Stathi, Χρύσανθος Νοταράς, Πατριάρχης Ιεροσολύμων, πρόδρομος του Νεοελληνικού Διαφωτισμού [Chrysanthos Notaras, Patriarch of Jerusalem, precursor of the Greek Enlightenment] (Athens: Syndesmos ton en Athinais Megaloscholiton, 1999). On Chrysanthos and science, see Nicolaos Kyriakos, "Chrysanthos Notaras o astronomos" (Chrysanthos Notaras the astronomer) (master's thesis, University of Thessalonica, 2007).

8. Stathi, Χρύσανθος Νοταράς, p. 110.

9. Ibid., p. 107.

10. The Greek title is Οδός μαθηματικής (Venice, 1749; 2nd ed., 1775).

11. On Anthrakites' trial, see Alkis Aggelou, "Η δίκη του Μεθοδίου Ανθρακίτη" [The trial of Methodios Anthrakites], in Των φώτων. Όψεις του νεοελληνικού διαφωτισμού [Lights: Aspects of Modern Greek Enlightenment] (Athens: Hermes, 1988), pp. 23–37.

12. On Poleni's teaching laboratory, see Antonio Salandin and Maria Pancino, *Il teatro di filosofia sperimentale di Giovanni Potent* (Padua: Lint editoriale associati, 1986).

13. Printed in Vienna in 1805. This Jesuit textbook was widely used in Europe

during the seventeenth and eighteenth centuries, after the improved editions done by Whiston (1710), Musschenbroek (1724), and Boscovich (1745).

14. This was a translation of a portion of the *Oeuvre* of Segner *Cursus mathematicus*, published in Halle in five volumes between 1758 and 1767.

15. See Stephen K. Batalden, *Catherine II's Greek Prelate Eugenios Voulgaris in Russia, 1771–1806* (New York: Columbia University Press, 1982).

16. Eugenios Voulgaris, *Α. Τακουετίου, Στοιχεία Γεωμετρίας* [Elements of geometry of A. Tacquet] (Vienna, 1805), pp. XI–XII.

17. Fortunatus a Brixia, *Philosophia sensuum mechanica methodice tractata* (Brescia, 1747).

18. Greek title, *Εισηγήσεις της εκλεκτικής φυσικής φιλοσοφίας*. This work by Voulgaris was never published; it circulated in schools in manuscript form, of which seventeen have been conserved. Wucherer was also author of al *Historia creationis, quatenus illa capite primo Geneseos* (Jena, 1729).

19. Nikephoros Theotokis, *Στοιχεία φυσικής εκ των νεωτέρων συνερανισθέντα* [Elements of physics compiled from the moderns], 2 vols. (Leipzig, 1766 and 1767).

20. On Theotokis, see George Vlahakis, "Nikiphoros Theotokis, Scientist and Theologian," in Graham Speake, ed., *Encyclopedia of Greece and the Hellenic Tradition*, vol. 2 (London: Fitzroy Dearborn, 2000), pp. 1673–74.

21. Paschalis Kitromilides, *The Enlightenment as Social Criticism: Iosipos Moisiodax and Greek Culture in the Eighteenth Century* (Princeton: Princeton University Press, 1992), p. 22. The student was Diamant Corayan, an important participant in the Modern Greek Enlightenment.

22. Iosipos Moisiodax, *Απολογία* [Apology] (Vienna, 1780), p. 166.

23. *Život i priključenija D. Obranović* (Leipzig, 1783, 1788), translated by G. R. Noyes as *The Life and Adventures of Dimitrije Obranović* (Berkeley and Los Angeles: University of California Press, 1953).

24. Kitromilides, *The Enlightenment as Social Criticism*, p. 25.

25. Ibid., p. 33.

26. Ibid., p. 38. Only Orthodox clergymen who had taken vows of chastity could obtain elevated posts. Married clergy were usually village priests.

27. Ibid., p. 65.

28. In fact, they had in mind the *Liber Abacci*, which gave methods for calculations. Use of these methods in the Orthodox world was delayed because of the tradition of Greek numbering. However, the first Greek abacus to be published was very successful; see Glyzounis's *Βιβλίον πρόχειρον τοις πάσι περιέχον την τε πρακτικήν αριθμητικήν...* [Handy book for all, containing practical arithmetic...] (Venice, 1568). This is the book to which Moisiodax's adversaries referred.

29. I. Moisiodax, *Θεωρία γεωγραφίας* [Theory of geography] (Vienna, 1781).

30. Ibid.

31. Kodrikas had the edition of the *Plurality of Worlds* that is found in the *Œuvres de monsieur de Fontenelle* (Paris, 1766).

32. On the relations between science and the Enlightenment, see Efthymios Nico-

Attica, that during the Miocene period a form existed there, which connected Semnopithecus and Macacus; and this probably illustrates the manner in which the other and higher groups were once blended together."

5. Costas B. Krimbas, "Αντιθέσεις ξένων και Ελλήνων για παλαιοντολογικές ανασκαφές τον 19ο αιώνα στην Ελλάδα" [Oppositions between foreigners and Greeks over the paleontological digs in the 19th century in Greece], in *Η επιστημονική σκέψη στον ελληνικό χώρο, 18ος–19ος αι* [Scientific thought in Greek space, 18th–19th c.] (Athens: Trochalia, 1998), pp. 195–200.

6. Sotiriadou, "Η εμφάνιση της θεωρίας της εξέλιξης των ειδών," p. 100.

7. S. Sougras, *Η νεωτάτη του υλισμού φάσις, ήτοι ο Δαρουϊνισμός και το ανυπόστατον αυτού* (Athens, 1876).

8. Ibid., p. 179.

9. This biography was based on the book by Wilhelm Preyer, *Darwin: Sein Leben und Wirken* (Berlin: Hoffmann, 1869).

10. Sotiriadou, "Η εμφάνιση της θεωρίας της εξέλιξης των ειδών," p. 108. *The Human Species* was published in 1877.

11. *Ο άνθρωπος και ο υλισμός* [Man and materialism] (Milan, 1882); *Ψυχολογικαί μελέται* [Psychological studies] (1887); *Περί γενέσεως του ανθρώπου* [On the Birth of Man] (1893); *Θρησκεία και επιστήμη* [Religion and science] (Trieste, 1894); and articles in the journal *Anaplasis* and the Greek newspaper of Trieste, *Ημέρα* [The day].

12. Skaltsounis, *Θρησκεία και επιστήμη*, pp. 207–208.

13. On Haeckel in Greece and the reaction of the church, see Kyriakos Kyriakou and Constantine Skordoulis, "The Reception of Ernst Haeckel's Ideas in Greece," *Almagest* 2 (2010): 84–103.

14. *Η κατά(ε)δάφισις του ελληνικού Πανεπιστημίου* [The demolition of the Greek university] (Leipzig, 1894).

15. Philopoimen Stephanides, *Αι υποθέσεις του υλισμού και του δαρβινισμού και το Ελληνικόν Πανεπιστήμιον* (April 1895).

16. Sotiriadou, "Η εμφάνιση της θεωρίας της εξέλιξης των ειδών," p. 132.

17. *Πρακτικά Συγκλήτου 1878–1880* [Minutes of the University Senate, 1878–1880], vol. 12.

18. Sotiriadou, "Η εμφάνιση της θεωρίας της εξέλιξης των ειδών," p. 142.

19. *Anaplasis*, no. 119 (1893): 1712.

20. *Anaplasis*, no. 62 (1890): 787.

21. Sotiriadou, "Η εμφάνιση της θεωρίας της εξέλιξης των ειδών," pp. 159–60.

22. A. Kousidis, *Θρησκεία, επιστήμη, κοινωνία* [Religion, science, society] (Athens, 1935), p. 68.

23. *Διακήρυξις της Χριστιανικής Ενώσεως Επιστημόνων* [Declaration of the Christian Union of Scientists] (Athens, 1946), p. 15.

24. Ibid., pp. 18, 20.

25. Ibid., pp. 39–41.

26. Ibid., pp. 43–46.

27. Ibid., p. 51.

28. Ibid., pp. 55, 92.

29. Speech by Costas B. Krimbas to the Academy of Athens on the occasion of the Year of Darwin, 17 February 2009.

30. Costas B. Krimbas, "The Evolutionary Worldview of Theodosius Dobzhansky," in Mark B. Adams, ed., *The Evolution of Theodosius Dobzhansky: Essays on His Life and Thought in Russia and America* (Princeton: Princeton University Press, 1994), pp. 179–93. See also Francisco Ayala, "'Nothing in Biology Makes Sense Except in the Light of Evolution': Theodosius Dobzhanski, 1900–1975," *Journal of Heredity* 68 (1977): 3–10.

31. Kantiotis, metropolitan of Florina, encyclical 403/13-3-1985.

32. Kallinikos, *Ο άνθρωπος από τον πίθηκο; Απάντηση στην υλιστική άποψη* [Man from monkey? Reply to the materialist thesis] (Athens, 1987).

33. Jon D. Miller, Eugienie C. Scott, and Shinji Okamoto, "Public Acceptance of Evolution," *Science* 313 (2006): 765–66; "Creation/Evolution among Eastern Orthodox Laity," 24 October 2008, posted on the Web site of the National Center for Science Education, http://ncse.com/news/2008. On the recent history of creationism in eastern Europe, see Ronald L. Numbers, *The Creationists: From Scientific Creationism to Intelligent Design*, expanded ed. (Cambridge, Mass.: Harvard University Press, 2006), pp. 412–16. A reanalysis of the statistics published in 2006 showed that Greece would have dropped even further if the graph had plotted "rejection of evolution"; see Mike Dickison, "Pictures of Numbers: Adding Variables," 14 August 2006, at www.numberpix.com/2008/08/adding_variables.html.

Aggelou, Alkis. "Η δίκη του Μεθοδίου Ανθρακίτη" [The trial of Methodios Anthrakites]. In *Των φώτων. Όψεις του νεοελληνικού διαφωτισμού* [Lights: Aspects of Modern Greek Enlightenment], pp. 23–37. Athens: Hermes, 1988.

Allard, André. *Maxime Planude: Grand calcul selon les Indiens.* Travaux de la Faculté de philosophie et lettres de l'Université catholique de Louvain, 27. 1981.

Anglod, Michael, ed. *The Cambridge History of Christianity.* Vol. 5, *Eastern Christianity.* Cambridge: Cambridge University Press, 2006.

Arabatzi, Aikaterini. "Η κοσμολογία του Ιωάννη Δαμασκηνού" [The cosmology of John of Damascus]. Master's thesis, University of Athens and National Technical University of Athens, 1999.

Arseniev, J. *Puteshestvye cherez Sibir ot Tobols'ka do Nerchinski i granits Kitaya . . .* [Voyage through Siberia from Tobolsk to Nershinksi and the Chinese border . . .]. St. Petersburg, 1882.

Arvanitakis, G. L. "Notes astronomiques." *Le Messager d'Athènes,* no. 5217 (February 1939).

Baddeley, J. F. *Russia, Mongolia, China, Being Some Record of the Relations between Them from the Beginning of the XVIIth Century to the Death of the Tsar Alexei Michailovich, A.D. 1602–1676 . . .* London: Macmillan, 1919.

Basil of Caesarea. *Homélies sur l'Hexaéméron.* Ed. Stanislas Giet. Sources Chrétiennes 26bis. Paris, 1968.

———. *Homilies on the Hexameron.* Trans. Sister Agnes Clare Way. Washington, D.C.: Catholic University of America Press, 1963.

Bianchi, Guido. "Sulla cultura astronomica di Giorgio di Pisidia." *Aevum* 40 (1966): 35–52.

Blay, Michel, and Efthymios Nicolaidis, eds. *L'Europe des sciences.* Paris: Seuil, 2001.

Blemmydes, Nicephorus. *Επιτομή φυσικής* [Epitome of physics]. Leipzig, 1784.

Boissonade, Jean-François. *Michiael Psellus: De operatione daemonum.* Nuremberg, 1838.

Bouché-Leclercq, Auguste. *L'astrologie grecque* [Greek astrology]. Paris, 1899.

Bouzoudi, Ioanna. "Nicephorus Blemmydes and His Treatise of Physics (1260): A Study of His Theory of Heaven and Its Sources." Master's thesis, Université de Lille III, 2000.

Browning, Robert. *Church, State and Learning in Twelfth Century Byzantium*. London: Dr Williams's Trust, 1881.

Byden, Börje. "Theodore Metochites Stoicheiosis Astronomike and the Study of Philosophy in Early Palaiologan Byzantium." PhD dissertation, Göteborg University, 2001.

Cacouros, Michel, and Marie-Hélène Congourdeau, eds. *Philosophie et sciences à Byzance de 1204 à 1453*. Leuven: Uitgeverij Peeters en Departement Oosterse Studies, 2006.

Callahan, John F. "Greek Philosophy and the Cappadocian Cosmology." *Dumbarton Oaks Papers* 12 (1958): 31–55.

Caudano, Anne-Laurence. *"Let There Be Lights in the Firmament of Heaven": Cosmological Depictions in Early Rus*. Palaeoslavica, 14, Supplementum 2. Cambridge, Mass., 2006.

———. "Un univers sphérique ou voûté? Survivance de la cosmologie antiochienne à Byzance (XI et XII s.)." *Byzantion* 78 (2008): 66–86.

Chatzis, Konstantinos. "La modernisation technique de la Grèce, de l'indépendance aux années de l'entre-deux-guerres: faits et problèmes d'interprétation." *Etudes Balkaniques* 3 (2004): 3–23.

Chrissidis, Nikolaos A. "Creating the New Educated Elite: Learning and Faith in Moscow's Slavo-Greco-Latin Academy, 1685–1694." PhD dissertation, Yale University, 2000.

———. "A Jesuit-Aristotle in Seventeenth-Century Russia: Cosmology and the Planetary System in the Slavo-Greco-Latin Academy." In Jarmo Kotilaine and Marshall Poe, eds., *Modernizing Muscovy: Reform and Social Change in Seventeenth-Century Russia*, pp. 391–416. London: Routledge Curzon, 2004.

Clucas, Lowell. *The Trial of John Italos and the Crisis of Intellectual Values in Byzantium in the Eleventh Century*. Munich: Institut für Byzantinistik der Universität München, 1981.

Comnena, Anna. *The Alexiad of Anna Comnena*. Ed. and trans. E. R. A. Sewter. Harmondsworth: Penguin, 1969.

Constantinides, Costas N. *Higher Education in Byzantium in the 13th and Early 14th Centuries*. Nicosia: Cyprus Research Centre, 1982.

Corydaleus, Theophilus. Εἴσοδος φυσικῆς ἀκροάσεως κατ'Ἀριστοτέλην [Introduction to the physics of Aristotle]. Venice, 1779.

———. Γενέσεως καὶ φθορᾶς κατ'Ἀριστοτέλην [On genesis and corruption according to Aristotle]. Venice, 1780.

Cumont, Franz, ed. *Catalogus Codicum astrologorum graecorum*. Vol. V 1. Brussels, 1904.

Declaration of the Christian Union of Scientists [Διακήρυξις τῆς Χριστιανικῆς Ἑνώσεως Ἐπιστημόνων]. Athens, 1946.

Delatte, Armand. *Anecdota Atheniensia et alia II: textes relatifs à l'histoire des sciences*. Paris: E. Droz, 1939.

Dialetis, Dimitris, Gavroglu Kostas, and Patiniotis Manolis. "The Sciences in the Greek Speaking Regions during the 17th and 18th Centuries." In Gavroglu Kostas, ed., *The Sciences at the Periphery of Europe during the 18th Century*, New Studies in the History and Philosophy of Science and Technology. Archimedes, vol. 2, pp. 41–71. Dordrecht: Kluwer Academic Publishers, 1999.

Duhem, Pierre. *Le système du monde*. Vol. 2. Paris: Hermann, 1914; 2nd ed., 1974.

Dzielska, Maria. *Hypatia of Alexandria*. Trans. F. Lyra. Cambridge, Mass.: Harvard University Press, 1996.

Evaggelatou-Notara, Fl. "Οποίον εστί μέρος της αστρολογίας κακιζόμενόν τε και αποτρόπαιον (Αστρολογία-Αστρονομία και οι σχετικές αντιλήψεις κατά τον IB΄αιώνα)" [Astrology-astronomy and related concepts during the 12th c.]. In Nikos Oikonomides, ed., *Το Βυζάντιο κατά τον 120 αιώνα. Κανονικό Δίκαιο, κράτος και κοινωνία* [Byzantium during the 12th c.: Canonical law, state and society], pp. 447–63. Athens: Etaireia Vyzantinon kai Metavyzantinon Meleton, 1991.

Ferngren, Gary, ed. *Science and Religion: A Historical Introduction*. Baltimore: Johns Hopkins University Press, 2002.

Firmin-Didot, Ambroise. *Alde Manuce et l'hellénisme à Venise*. Paris, 1875; repr., Brussels: Culture et Civilisation, 1966.

Fonkich, Boris. *Grechko-russkie kulturnye sviazi v XV–XVII vv. Grecheshkie rukopisi v Rosii* [Greek-Russian cultural relations in 15th–17th c. Greek manuscripts in Russia]. Moscow, 1972.

Gadolin, Anitra. *A Theory of History and Society with Special Reference to the Chronographia of Michael Psellus: 11th Century Byzantium and a Related Section on Islamic Ethics*. Amsterdam: A. M. Hakkert, 1987.

Gautier, Paul, ed. *Michel Italikos, Lettres et discours*. Archives de L'Orient Chrétien 14. Paris: Institut Français d'Études Byzantines, 1972.

Glyzounis, Emmanuel. *Βιβλίον πρόχειρον τοις πάσι περιέχον την τε πρακτικήν αριθμητικήν . . .* [Handy book for all, containing practical arithmetic . . .]. Venice, 1568.

Golvers, Noël, and Efthymios Nicolaidis. *Ferninand Verbiest and Jesuit Science in China: An Annotated Edition and Translation of the Constantinople Manuscript (1676)*. Athens: NHRF–F. Verbiest Institute, 2009.

Gregoras, Nikephorus. *Byzantina historia*. Ed. L. Schopen. Corpus scriptorium histariae byzantinae VIII. Bonn, 1829.

Gregory of Nyssa. *Contre le destin*. In *Les Pères de l'église et l'astrologie*, Introduction by Marie-Elisabeth Allamany, ed. J.-P. Migne. Les pères dans la foi. Paris: J.-P. Migne, 2003.

———. *Explicatio apologetica ad Petrum fratrem, in hexaemeron* [Apologia to his brother Peter on the Hexaemreon]. Patrologia Graeca, 44. Ed. J.-P. Migne: Paris, 1857–66.

Gregory Palamas. *Κεφάλαια εκατόν πεντήκοντα, φυσικά και θεολογικά, ηθικά τε και πρακτικά και καθαρτικά της βαρλααμίτιδος λύμης* [One hundred and fifty chapters, physical and theological, moral and practical and purifying against the Barlaamite sickness]. In *Complete Works of Gregory Palamas*, vol. 8, ed. P. K. Christou. Thessalonica: Patristic Editions Gregory Palamas, 1994.

———. *Λόγος αντιρρητικός προς Ακίνδυνον* [Refutation of Akindynos]. In *Complete*

Works of Gregory Palamas, vol. 6, ed. P. K. Christou. Thessalonica: Patristic editions Gregory Palamas, 1987.

Grosdidier De Matons, J. "Les thèmes d'édification dans la Vie d'André Salos." *Travaux et Memoires* 4 (1970): 277–328.

Grumel, Venance. *La Chronologie*. Traités d'Études byzantines I. Paris: Presses Universitaires France, 1958.

Haskins, Charles H. *The Renaissance of the Twelfth Century*. Cambridge, Mass.: Harvard University Press, 1927.

Hering, Gunnar. *Das ökumenische Patriarchat und europäische Politik 1620–1638*. Wiesbaden: Franz Steiner, 1968.

Hill-Paulus, Beate. *Nikolaj Gavrilovic Spatharij (1636–1708) und seine Gesandtschaft nach China*. Hamburg-Tokyo: Gessellschaft für Natur und Völkerkunde Ostasiens, 1978.

Hionides, Harry T. *Paisius Ligarides*. New York: Twayne, 1972.

Hunger, H. *Schreiben und Lesen in Byzanz: Die byzantinische Buchcultur*. Munich: Verlag C. H. Beck, 1989.

Iliou, Philippos. *Τύφλωσον κύριε τον λαόν σου: Οι προεπαναστατικές κρίσεις και ο Νικόλαος Πίκκολος* [God, blind your people: Prerevolutionary crisis and Nicolas Piccolo]. Athens: Poreia, 1988.

Inglebert, Hervé. *Interpretatio Christiana. Les mutations des savoirs (cosmographie, géographie, ethnographie, histoire) dans l'Antiquité chrétienne (30–630 après J.-C.)*. Etudes Augustiniennes. Série Antiquité 166. Paris, 2001.

Iorga, Nicolae. *Byzance après Byzance*. Bucharest, 1935; repr., Paris: Balland, 1992.

John Chrysostom. *Homilies on Genesis* (1–67 in 3 vols.). Trans. Robert C. Hill. Washington, D.C.: Catholic University of America Press, 1986.

———. *Υπόμνημα εις την Γένεσιν* [Memorandum on Genesis], *Homilies I–XXIII*. Ed. Spyridon Moustakas. Thessalonica: Patristic editions Gregory Palamas 1981.

John of Damascus. *De fide ortodoxa lib II*. Patrologia Graeca, 94.

———. *De imaginibus oratio I*. Patrologia Graeca, 94.

———. "De imaginibus oratio I." In *Writings: The Fathers of the Church*, vol. 37, trans. Fredereick Chase. Washington, D.C.: Catholic University Press of America, 1979.

Jones, Alexander. *An Eleventh-Century Manual of Arabo-Byzantine Astronomy*. Corpus des Astronomes Byzantins III. Amsterdam: J. C. Gieben, 1987.

Kallinikos (Metropolites of Piraeus). *Ο άνθρωπος από τον πίθηκο; Απάντηση στην υλιστική άποψη* [Man from monkey? Reply to the materialist thesis]. Athens, 1987.

Karas, Yannis. *Οι επιστήμες στην Τουρκοκρατία: Χειρόγραφα και έντυπα* [Sciences under Ottoman domination: Printed books and manuscripts]. Vol. 1, *Mathematics*; vol. 2, *Natural Sciences*; vol. 3, *Life Sciences*. Athens: Estia, 1992–93.

———. *Οι θετικές επιστήμες στον ελληνικό χώρο, 15ος–19ος αιώνας* [Exact sciences in the Greek world, 15th–19th c.]. Athens: Daidalos / Zacharopoulos, 1991.

———, ed. *Ιστορία και φιλοσοφία των επιστημών στον ελληνικό χώρο (17ος–19ος αι.)* [History and philosophy of science in the Greek world, 17th–19th c.]. Athens: Metaichmio, 2003.

Kartsonakis, Manolis. "Η σύνοψις των Φυσικών του Συμεών Σηθ" [On natural things of Symeon Seth]. In George Vlahakis and Efthymios Nicolaidis, eds., *Βυζάντιο-Βενετία-Νεώτερος ελληνισμός* [Byzantium-Venice-Modern Hellenism], pp. 129–37. Athens: NHRF, 2004.

Katsiampoura, Gianna. "Comparing Two Byzantine *Quadrivia*: The *Quadrivium* of 1008 and G. Pachymeres Syntagma. Resemblances and Differences." In *Atti del Convegno "Libri di Scuola e Pratiche Didattiche dall'Antichita al rinancimento,"* pp. 410–25. Cassino: Eitioni Università degli Studi di Cassino, 2010.

———. "John Grammatikos, Scientist and/or Magus?" [in Greek]. In K. Skordoulis et al., eds., *Ζητήματα επιστήμης: ιστορία, φιλοσοφία και διδακτική* [Questions of science: History, philosophy and didactics], pp. 29–36. Athens: Nissos, 2008.

———. "Nikephoros Gregoras versus Barlaam of Calabria : A Debate over Predicting Eclipses in Constantinople in 14th c." [in Greek]. *Neusis* 13 (2004): 138–48.

———. "Reception, Diffusion and Functioning of Science in Middle Byzantium and the *Quadrivium* of 1008" [in Greek]. PhD dissertation, University of Panteion, Athens, 2004.

———. "Transmutation of Matter in Byzantium: The Case of Michael Psellos, the Alchemist." *Science and Education* 17 (2008): 663–68.

Katsiavrias, Nicolaos. "Η κοσμοαντίληψη του Αγίου Γρηγορίου του Παλαμά (1296–1359)" [The perception of the world of Saint Gregory Palamas, 1296–1359]. PhD dissertation, University of Athens, 2001.

Kazhdan, Alexander, with L. E. Sherry and Chr. Angelidi. *A History of Byzantine Literature (650–850)*. Athens: Institute for Byzantine Research / NHRF, 1999.

Kedrenos,Georgios. *Σύνοψις Ιστοριών* [Summary of Histories]. Ed. I. Bekker. Vol. II. Bonn, 1838–39.

Kitromilides, Paschalis. *The Enlightenment as Social Criticism: Iosipos Moisiodax and Greek Culture in the Eighteenth Century*. Princeton: Princeton University Press, 1992.

Kousidis, Aristippos. *Θρησκεία, επιστήμη, κοινωνία* [Religion, science, society]. Athens, 1935.

Krimbas, Costas B. "Alexandre Theotokis, la notion de l'évolution et le premier texte de zoologie grecque." *Historical Review*, Institute for Neohellenic Research, 4 (2007): 191–97.

———. "Αντιθέσεις ξένων και Ελλήνων για παλαιοντολογικές ανασκαφές τον 19ο αιώνα στην Ελλάδα" [Oppositions between foreigners and Greeks over the paleontological digs in 19th c. in Greece]. In *Η επιστημονική σκέψη στον ελληνικό χώρο, 18ος–19ος αι.* [Scientific thought in Greek space, 18th–19th c.], pp. 195–200. Athens: Trochalia, 1998.

Kritikos Theodoros. *Η πρόσληψη της επιστημονικής σκέψης στην Ελλάδα. Η Φυσική μέσα από πρόσωπα, θεσμούς και ιδέες* [The reception of scientific thought in Greece: Physics through persons, institutions and ideas]. Athens: Papazisis, 1995.

Kunitzsch, Paul. "Die arabische Herkunft von zwei Sternverzeichnissen in Cod. Vat. Gr. 1056." *Zeitschrift der Deutschen Morgenländ* 120 (1970): 281–87.

Kurtz, Eduard, and Franz Drexl. *Michaellis Pselli. Scripta minora*. Vol. I, Milan: Societa editrice Vita e Pensiero, 1936.

Kyriakos, Nicolaos. "Χρύσανθος Νοταράς ο Αστρονόμος" [Chrysanthos Notaras the astronomer]. Master's thesis, University of Thessalonica, 2007.

Kyriakou, Kyriakos, and Constantine Skordoulis. "The Reception of Ernst Haeckel's Ideas in Greece." *Almagest* 2 (2010): 84–103.

Lampakis, Stylianos. *Γεώργιος Παχυμέρης, Πρωτέκδικος και Διακαιοφύλαξ* [Georgios Pachymeris, Protekdikos and Dikaiophylax]. Athens: NHRF, 2004.

Legrand, Emile. *Epistolaire grec ou recueil des lettres adressées par la plupart à Chrysante Notaras.* Paris, 1888.

Leloup, Jean-Yves. *Écrits sur l'hésychasme.* Paris: Albin Michel,1999.

Lemerle, Paul. *Le premier humanisme byzantin.* Paris: PUF, 1971.

Les Pères de l'église et l'astrologie [Church Fathers and Astrology]. Introduction by Marie-Elisabeth Allamany, ed. J.-P. Migne. Les pères dans la foi. Paris: J.-P. Migne, 2003.

Leurquin, Régine. *Théodore Méliténiote, Tribiblos astronomique, livre I.* Corpus des astronomes byzantins IV. Amsterdam, 1990. *Livre II*, Corpus des astronomes byzantins V–VI. Amsterdam, 1993.

Lindberg, David C. "Science and the Early Church." In David C. Lindberg and Ronald L. Numbers, eds., *God and Nature: Historical Essays on the Encounter between Christianity and Science*, pp. 19–48. Berkeley and Los Angeles: University of California Press, 1986.

Lindberg, David C., and Ronald L. Numbers, eds. *When Science and Christianity Meet.* Chicago: University of Chicago Press, 2003.

Magdalino, Paul. "The Byzantine Reception of Classical Astrology." In Catherine Holmes and Judith Waring, eds., *Literacy, Education and Manuscript Transmission in Byzantium and Beyond*, pp. 33–57. Leiden: Brill, 2002.

———. *L'orthodoxie des astrologues. La science entre le dogme et la divination à Byzance, VIIe–XIVe siècle.* Réalités byzantines 12. Paris: Lethielleux, 2006.

Magdalino, Paul, and Maria Mavroudi, eds. *The Occult Sciences in Byzantium.* Geneva: La pomme d'or, 2006.

Makrides, Vasilios. *Die religiöse Kritik am kopernikanischen Weltbild in Griechenland zwischen 1794 und 1821: Aspekte griechisch-orthodoxer Apologetik angesichts naturwissenschaftlicher Fortschritte.* Frankfurt am Main: Peter Lang Verlag, 1995.

Matalas, Paraskevas. *Εθνος και Ορθοδοξία. Από το «ελλαδικό» στο βουλγαρικό σχήσμα* [Nation and Orthodoxy: From Greek to Bulgarian schism]. Herakleion: Crete University Press, 2003.

Mathiesen, Robert. "Magic in Slavia Orthodoxa: The Written Tradition." In Henry Maguire, ed., *Byzantine Magic*, pp. 155–78. Washington, D.C.: Dumbarton Oaks Research Library and Collection, 1995.

Mavrocordato, Alexandro. *Pneumaticum instrumentum circulandi sanguinis sive de motum et usu pulmonum.* Bononiae, 1664.

Maximus the Confessor. *Capita Alia.* Patrologia Graeca, 90.

———. *Epistole.* Patrologia Graeca, 91.

Mendieta, E. Amand de. "Les neuf Homélies de Basile de Césarée sur l'Hexaéméron." *Byzantion* 48 (1978): 337–68.

Mercier, Raymond. *An Almanac for Trebizond for the Year 1336*. Corpus des Astronomes Byzantins VII. Louvain-la-Neuve, 1994.

———. "The Greek *Persian Syntaxis* and the *Zīj-i Īlkhānī*." *Archives Internationales d'Histoire des Sciences* 35 (1985): 436–38.

———. "The Parameters of the Zīj of Ibn al-'Alam." *Archives Internationales d'Histoire des Sciences* 39 (1989): 22–50.

Meredith, Anthony. *The Cappadocians*. Crestwood, N.Y.: St. Vladimir's Seminar Press, 1995.

Mergiali, Sophia. *L'enseignement et les lettrés pendant l'époque des Paléologues*. Athens: Etaireia ton Filon tou Laou, 1996.

Methodios Anthrakites. Οδός μαθηματικής [Course of mathematics]. Venice, 1749.

Meyendorff, John. "Les débuts de la controverse hésychaste." *Byzantion* 23 (1953): 87–120.

———. "Un mauvais théologien de l'unité au XIVe siècle: Barlaam le Calabrais." In *L'Église et les Églises, 1054–1954: Etudes et travaux offerts a Dom Lambert Baudouin*, II, pp. 47–64. Chévetogne: Éd. De Chévetogne, 1954.

———. *St. Gregory Palamas and Orthodox Spirituality*. Trans. Adele Fiske. New York: St. Vladimir's Seminary Press, 1974.

Moffatt, Ann. "Schooling in the Iconoclast Centuries." In A. A. M. Bryer and J. Herrin, eds., *Iconoclasm*, pp. 85–92. Birmingham: Center for Byzantine Studies, 1977.

Mogenet, Joseph, A. Tihon, R. Royez, and A. Berg. *Nicéphore Grégoras, calcul de l'éclipse de soleil du 16 juillet 1330*. Corpus des astronomes byzantins I. Amsterdam, 1983.

Neugebauer, Otto. *A History of Ancient Mathematical Astronomy*. 3 vols. Berlin: Springer-Verlag, 1975.

Nicolaidis, Efthymios. "Les élèves grecs de l'Ecole polytechnique (1820–1921)." In *La diaspora hellénique en France*, pp. 55–65. Athens: École française d'Athènes, 2000.

———. "The Spread of New Science to Southeastern Europe: During or before the Greek Enlightenment?" In Celina Lertora Mendoza, E. Nicolaidis, and Jan Vandersmissen, eds., *The Spread of the Scientific Revolution in the European Periphery, Latin America and East Asia*, Proceedings of the XXth International Congress of History of Science, vol. V, pp. 33–45. Turnhout, Belgium: Brepols, 1999.

———. "Was the Greek Enlightenment a Vehicle for the Ideas of the Scientific Revolution?" *Balkan Studies* 40 (1999): 7–19.

Noica, Constantin. "La signification historique de l'œuvre de Théophile Corydalée." *Revue des Etudes de Sud-Est Européennes* 2 (1973): 285–306.

Nomikou, Christine, and Gianna Katsiampoura. "The School of Magnavra: Sciences in Byzantium" [in Greek]. In K. Skordoulis et al., eds., Ζητήματα επιστήμης: Ιστορία, φιλοσοφία και διδακτική [Questions of science: History, philosophy and didactics], pp. 37–43. Athens: Nissos, 2008.

Notaras, Chrysanthos. Εισαγωγή εις τα γεωγραφικά και σφαιρικά [Introduction to geography and to the sphere]. Paris, 1716.

Obolensky, Dimitri. *The Byzantine Commonwealth: Eastern Europe, 500–1453*. New York: Praeger, 1971.

Obranovič, D. *Život i priključenija D. Obranovič.* Leipzig, 1783, 1788. Translated by
G. R. Noyes as *The Life and Adventures of Dimitrije Obranovič.* Berkeley and Los
Angeles: University of California Press, 1953.

Pachymeres, George. *Traité des quatre leçons, arithmétique, musique, géométrie et
astronomie.* Ed. P. Tannery and E. Stéphanou. Studi e Testi 94. Rome: Vatican,
1940.

Palmer, W. *The Patriarch and the Tsar,* vol. III. London: Trubner, 1873.

Panaitescu, P. "Nicolas Spathar Milescu." *Mélanges de l'École Roumaine en France* 1
(1925): 33–180.

Papadopoulos, Chrysostomos. *Οι πατριάρχαι Ιεροσολύμων ως πνευματικοί χειραγωγοί της Ρωσσίας κατά τον ΙΖ΄ αιώνα* [The patriarchs of Jerusalem: Spiritual guides
of Russia in the 17th c.]. Jerusalem: Iero Koino Panagion Tafou, 1907.

Papadopoulos-Kerameus, Athanasios. *Ιεροσολυμιτική Βιβλιοθήκη* [Library of Jerusa-
lem]. Vol. IV. St. Petersburg, 1899; repr., Brussels, 1963.

Paschos, Emmanuel A., and Panagiotis Sotiroudis. *The Schemata of the Stars.* Singapore: World Scientific Publishing, 1998.

Petsios, Kostas. *Η περί φύσεως συζήτηση στη νεοελληνική σκέψη* [The discussion on
nature in Modern Greek thought]. Jannina, 2003.

Phili, Christine. "Ioannis Carandinos (1784–1834). L'initiateur des mathématiques
françaises en Grèce." *Archives Internationales d' Histoire des Sciences* 56 (2006):
79–124.

Philo of Alexandria. *De Opificio Mundi,* "A Treatise on the Account of Creation of
the World." In *The Works of Philo Judeaus,* trans. Charles Duke Yonge. London:
H. G. Bohn, 1854–90.

———. *De Opificio Mundi.* Ed. Roger Arnaldez. Paris: Éditions du cerf 1961.

Philoponus, John. *Contra Proclum.* Ed. H. Rabe. Leipzig: Teubner, 1899.

———. *La Création du monde.* Trans. Marie-Claude Rosset and M. H. Congourdeau.
Paris: Migne, 2004.

Philotheos Kokkinos. *Λόγος εις τον εν αγίοις πατέρα ημών Γρηγόριον αρχιεπίσκοπον
Θεσσαλονίκης* [Discourse on our holy father Gregory, archbishop of Thessalo-
nica]. Patrologia Graeca, 51, cols. 551–656.

Photius. *Bibliothèque.* Ed. R. Henry. Vol. IV. Paris: Belles Lettres, 1963.

Picot, E. "Nicolas Spathar Milescu." In E. Legrand, ed., *Bibliographie hellénique,* 4:62–
104. Paris, 1896; repr., Brussels: Culture et Civilisation, 1963.

Pingree, David. "The Astrological School of John Abramius." *Dumbarton Oaks
Papers* 25 (1971): 191–215.

———. *The Astronomical Works of Gregory Chioniades.* Vol. 1, *The Zīj al'Alā'ī,* part 1:
Text, Translation, Commentary, part 2: *Tables,* Corpus des Astronomes Byzantins
II, 2 vols. Amsterdam, 1985–86.

———. "The Byzantine Version of Toledan Tables: The Work of George Lapithes?"
Dumbarton Oaks Papers 30 (1976): 87–132.

———. "Classical and Byzantine Astrology in Sassanian Persia." *Dumbarton Oaks
Papers* 43 (1989): 227–39.

———. "The Indian and Pseudo-Indian Passages in Greek and Latin Astronomical and Astrological Texts." *Viator* 7 (1976): 141–95.

Pisidae, Georgii. *Opus sex dierum, seu, mundi opificium*. Lutetia, 1584.

Preyer, Wilhelm. *Darwin: Sein Leben und Wirken*. Berlin: Hoffmann, 1869.

Psimmenos, Nikos. "Η 'εκ παλαιών και νεωτέρων ερανισθείσα' Προθεωρία της 'Εισόδου Φυσικής Ακροάσεως' του Θεοφίλου Κορυδαλέως" [The prologue of the *Introduction to Aristotle's Physics* of Th. Korydaleus]. *Πρακτικά Πανελλήνιου Επιστημονικού Συνεδρίου "Το αίτημα της διεπιστημονικής έρευνας. Οι επιστήμες στον ελληνικό χώρο"* ["The question of interdisciplinary research: Sciences in the Greek world," conference proceedings]. Athens, 1999.

Romanchuk, Robert. *Byzantine Hermeneutics and Pedagogy in the Russian North: Monks and Masters at the Kirillo-Belozerskii Monastery, 1397–1501*. Toronto: University of Toronto Press, 2007.

Rosenthal, Franz. "From Arabic Books and Manuscripts." *Journal of the American Oriental Society* 83 (1963): 452–57.

Rousseau, Philip. *Basil of Caesarea*. The Transformation of the Classical Heritage, 20. Berkeley: University of California Press, 1994.

Runia, David T. *Philo of Alexandria and the* Timaeus *of Plato*. Philosophia Antiqua 44. Leiden, 1986.

———. *Philo in Early Christian Literature: A Survey*. Compendia Rerum Iudaicarum ad Novum Testamentum III, vol. 3. Assen: Van Gorcum; Minneapolis: Fortress Press, 1993.

Ryan, William F. *Russian Magic at the British Library: Books, Manuscripts, Scholars, Travelers*. London: British Library, 2006.

Salandin, Antonio, and Pancino Maria. *Il teatro di filosofia sperimentale di Giovanni Potent*. Padua: Lint edotoriale associati, 1986.

Sambursky, Samuel. *The Physical World of Late Antiquity*. London: Routledge-Kegan, 1962.

Sangin, Mstislav Antonini. *Codices Rossicos*. Catalogus Codicum Astrologorum Graecorum, XII. Brussels, 1936.

Sathas, Konstantinos. *Μεσαιωνικη Βιβλιοθήκη* [Medieval Library]. Vol. 1. Athens, 1872.

Scholarios. *Oeuvres complètes*. Ed. L. Petit, A. Sideridès, and M. Jugie. Vol. IV. Paris: Maison de la bonne presse, 1931.

Sevčenko, Ihor. *Études sur la polémique entre Théodore Métochite et Nicéphore Choumnos*. Brussels: Editions de Byzantion, 1962.

———. "Remarks on the Diffusion of Byzantine Scientific and Pseudo-scientific Literature among the Orthodox Slavs." *Slavonic and East European Review* 59 (1981): 321–45.

Skaltsounis, Ioannis. *Ο άνθρωπος και ο υλισμός* [Man and materialism]. Milan, 1882.

———. *Θρησκεία και επιστήμη* [Religion and science]. Trieste, 1894.

———. *Περί γενέσεως του ανθρώπου* [On the birth of man]. Athens, 1893.

Sklavenitis, Triantafyllos, ed. *Τα ιδρυτικά κείμενα του Εθνικού Ιδρύματος Ερευνών και η αλληλογραφία Ι. Στ. Πεσμαζόγλου και Λ. Θ. Ζέρβα* [Founding documents of the National Hellenic Research Foundation and the correspondence of I. St. Pesmazoglou and L. Th. Zervas]. Athens: NHRF, 2008.

Smirnov, S. *Istoria Moskovskoj Slavjano-greko-latinskoj Akademmii* [History of the Slavo-Helleno-Latin Academy]. Moscow, 1855.

Solon, Peter. "The *Six Wings* of Immanuel Bonfils and Michael Chrysokokkes." *Centaurus* 15 (1970): 1–20.

Sorabji, Richard, ed. *Philoponus and the Rejection of Aristotelian Science.* London: Duckworth, 1987.

Sotiriadou, Anna. "Η εμφάνιση της θεωρίας της εξέλιξης των ειδών, δεδομένα από τον Ελληνικό χώρο" [The appearance of the theory of evolution of species, data from Greek space]. PhD dissertation, University of Thessalonica, 1990.

Sougras, Spyridon. *Η νεωτάτη του υλισμού φάσις, ήτοι ο Δαρουϊνισμός και το ανυπόστατον αυτού* [The most recent phase of materialism: Darwinism and its lack of foundation]. Athens, 1876.

Stathi, Pinelopi. *Χρύσανθος Νοταράς, Πατριάρχης Ιεροσολύμων, πρόδρομος του Νεοελληνικού Διαφωτισμού* [Chrysanthos Notaras, Patriarch of Jerusalem, precursor of the Greek Enlightenment]. Athens: Syndesmos ton en Athinais Megaloscholiton, 1999.

Stephanides, Philopoimen. *Αι υποθέσεις του υλισμού και του δαρβινισμού και το Ελληνικόν Πανεπιστήμιον* [The hypotheses of materialism and Darwinism and the Greek university]. Athens, April 1895.

Swerdlow, Noël M., and Otto Neugebauer. *Mathematical Astronomy in Copernicus' De Revolutionibus.* 2 vols. New York: Springer-Verlag, 1984.

Symeon, Seth. *Σύνοψις φυσικής* [On natural things]. Patrologia Graeca, 122, cols. 783–819.

Tannery, Paul. *Mémoires scientifiques*, XIII (Correspondence). Toulouse: Edouart Privat, 1934.

Tatakis, Basile. *La philosophie byzantine.* Paris: Presses Universitaires de France, 1949.

Tavardon, Paul. "Recherche sur l'astronomie byzantine; un aspect de la première renaissance des Paléologues." PhD dissertation, University of Toulouse, 1987.

Theotokis, Alexandre. *Γενικοί ζοολογικοί πίνακες ή Πρόδρομος της ελληνικής ζωολογίας* [General zoological tables, or forerunner to Greek zoology]. Corfu, 1848.

Theotokis, Nikephoros. *Στοιχεία φυσικής εκ των νεωτέρων συνερανισθέντα* [Elements of physics compiled from the moderns]. 2 vols. Leipzig, 1766 and 1767.

Tihon, Anne. "L'astronomie à Byzance à l'époque iconoclaste." In P. L. Butzer and D. Lohrmann, eds., *Science in Western and Eastern Civilization in Carolingian Times*, pp. 181–203. Basel: Birkhäuser, 1993.

———. "L'astronomie byzantine (du Ve au XVe siècle)." *Byzantion* 51 (1981): 603–24.

———. "L'astronomie byzantine à l'aube de la renaissance." *Byzantion* 66 (1996): 245–80.

———. "L'enseignement scientifique à Byzance." *Organon* 24 (1988): 89–108.

———. "Les tables astronomiques persanes à Constantinople dans la première moitié du XIVe siècle." *Byzantion* 57 (1987): 471–87.

———. "Sur l'identité de l'astronome Alim." *Archives Internationales d'Histoire des Sciences* 39 (1989): 3–21.

———. "Tables islamiques à Byzance." *Byzantion* 60 (1990): 401–25.

———. "Traités byzantins sur l'astrolabe." *Physis* 32 (1995): 323–57.

———. "Un traité astronomique chypriote du XIVe siècle. I." *Janus* 64 (1977): 279–308; II, 66 (1979): 49–81; III, 68 (1981): 65–127.

Tihon, Anne, and Raymond Mercier. *Georges Gémiste Pléthon Manuel d'Astronomie.* Corpus des astronomes byzantins IX. Louvain-la-Neuve, 1998.

Tsakoumis, Apostolos. "Τέσσερα ανέκδοτα κείμενα του Χρυσάνθου Νοταρά" [Four unpublished texts of Chrysanthos Notatars]. In E. Nicolaidis, ed., *Οι μαθηματικές επιστήμες στην Τουρκοκρατία* [Mathematical sciences during the Ottoman domination], pp. 147–222. Athens: INR/NHRF, 1992.

Tsourkas, Cléobule. *Les débuts de l'enseignement philosophique et de la libre pensée dans les Balkans. La vie et l'œuvre de Théophile Corydalée (1570–1646).* Bucarest, 1948; 2nd ed., Thessalonica: Institute for Balkan Studies, 1967.

University of Athens' Senate Minutes, 1878–1880 (Πρακτικά Συγκλήτου). Vol. 12.

Veselovsky, I. N. "Copernicus and Nasīr al-Dīn al-Tūsī." *Journal for the History of Astronomy* 4 (1973): 128–30.

Vlahakis, George. "Nikiphoros Theotokis, Scientist and Theologian." In Graham Speake, ed., *Encyclopedia of Greece and the Hellenic Tradition,* 2:1673–74. London: Fitzroy Dearborn, 2000.

Voulgaris, Eugenios. *A. Τακουετίου, Στοιχεία Γεωμετρίας* [Elements of geometry of A. Tacquet]. Vienna, 1805.

Walbridge, John. *The Wisdom of the Mystic East, Suhrawardi and Platonic Orientalism.* Albany: State University of New York Press, 2001.

Westerink, Leendert Gerrit. *Michael Psellus de omnifaria doctrina,* with a critical text and introduction. Utrecht: J. L. Beijers, 1948.

Wolska-Conus, Wanda. *Cosmas Indicopleustès, Topographie Chrétienne.* Sources Chrétiennes 141. Paris: Les Editions du Cerf , 1968.

Yalamas, Dimitris. "The Students of the Leichoudis Brothers at the Slavo-Greco-Latin Academy of Moscow." *Cyrillomethodianum* 15–16 (1991–92): 113–44.

Zervos, Christian. *Un philosophe néoplatonicien du XIe siècle: Michel Psellos, sa vie, son œuvre, ses luttes philosophiques, son influence.* Preface by François Picavet. Paris, 1920; New York: B. Franklin, 1973.